INTERNATIONAL SERIES IN

HEATING, VENTILATION AND REFRIGERATION

GENERAL EDITOR: N. S. BILLINGTON

VOLUME 15

AUTOMATIC CONTROLS FOR HEATING AND AIR CONDITIONING

Principles and Applications

Other Titles in the Series

CROOME
Noise, Buildings and People

CROOME & ROBERTS
Airconditioning and Ventilation of Buildings, 2nd Edition, Volume 1

DOWN
Heating and Cooling Load Calculations

KUT
Heating and Hot Water Services in Buildings
Warm Air Heating

MACKENZIE-KENNEDY
District Heating: Thermal Generation and Distribution

MEACOCK
Refrigeration Processes

OSBORNE
Fans, 2nd Edition

Related Pergamon Journals*

BUILDING AND ENVIRONMENT

ENERGY, the International Journal

ENERGY CONVERSION & MANAGEMENT

JOURNAL OF HEAT RECOVERY SYSTEMS

SOLAR ENERGY

**Free specimen copy sent on request*

AUTOMATIC CONTROLS FOR HEATING AND AIR CONDITIONING

Principles and Applications

by

K. M. LETHERMAN

BSc, MSc, PhD, CEng, MIEE, MCIBS

Department of Building, University of Manchester Institute of Science and Technology, England

PERGAMON PRESS

OXFORD · NEW YORK · TORONTO · SYDNEY · PARIS · FRANKFURT

U.K.	Pergamon Press Ltd., Headington Hill Hall, Oxford OX3 0BW, England
U.S.A.	Pergamon Press Inc., Maxwell House, Fairview Park, Elmsford, New York 10523, U.S.A.
CANADA	Pergamon Press Canada Ltd., Suite 104, 150 Consumers Rd., Willowdale, Ontario M2J 1P9, Canada
AUSTRALIA	Pergamon Press (Aust.) Pty. Ltd., P.O. Box 544, Potts Point, N.S.W. 2011, Australia
FRANCE	Pergamon Press SARL, 24 rue des Ecoles, 75240 Paris, Cedex 05, France
FEDERAL REPUBLIC OF GERMANY	Pergamon Press GmbH, 6242 Kronberg-Taunus, Hammerweg 6, Federal Republic of Germany

First edition 1981

British Library Cataloguing in Publication Data

Letherman, K. M.
Automatic controls for heating and air conditioning. - (International series in heating, ventilation and refrigeration; v. 15)
1. Heating - Control
2. Air conditioning - Control
I. Title II. Series
697 TH7226

ISBN 0-08-023222-1

Library of Congress Catalog Card no.: 80-42155

In order to make this volume available as economically and as rapidly as possible the author's typescript has been reproduced in its original form. This method unfortunately has its typographical limitations but it is hoped that they in no way distract the reader.

Printed in Great Britain by A. Wheaton & Co. Ltd., Exeter

To
ROSEMARY
SIMON and SOPHIE

PREFACE

This book has developed mainly out of the author's teaching of automatic control to the Master of Science course in Building Services Engineering at UMIST over the past 10 years. It is hoped that it will be of interest to those who wish to understand the relationships between theory and practice in automatic controls, although it is directed mainly at engineering-degree-level students, and at practitioners who wish to gain an understanding of control theory. The mathematical level assumes some familiarity with the ideas of calculus and complex numbers, although the matrix theory on which the more recent developments in multivariable control is based was felt to be beyond the scope of this text.

In order to control any physical process it is necessary to measure its behaviour, and Chapter 1 describes the range of sensors and methods for measurement of the two variables mainly of interest in building services systems, temperature and humidity.

Chapter 2 consists of a series of notes on the application of controls to a number of specific areas of building environmental services, with a brief description of the dynamic properties of some of the components of controlled systems.

In Chapter 3 the detailed mechanisms and circuits of pneumatic, electric and electronic controller types are explained. Their general characteristics are described in Chapter 4.

Chapter 5 presents the basic ideas upon which the theory of linear automatic control is based. The mathematical description of system dynamics by Laplace transfer functions is introduced, and it is shown how the transient and frequency response behaviour of a system can be calculated. The standard methods of stability prediction are described, with worked examples. Since many practical systems are in fact non-linear, Chapter 6 deals with some of the methods of analysis available for non-linear cases.

A number of problems are given in addition to the worked examples in the text. The problems marked UMIST have appeared in final degree examination papers and are reproduced by kind permission of the Registrar.

A fairly extensive bibliography is provided so that those who wish to do so may pursue aspects of the subject in further detail. In the Appendix some transform results are derived which are useful in characterising the experimentally measured step responses of practical systems.

In a work such as this some errors are perhaps inevitable in spite of ones best efforts. The author would be grateful to be informed of any errors which readers may find, and for these he takes full responsibility.

Manchester
May 1980

CONTENTS

LIST OF SYMBOLS

Symbol	Meaning	Units	Section of first appearance
A	Gain factor	$^{o}C/W$	5.4.2
A	Thermistor equation coefficient	ohms	1.1.3
A	Numerical constant	-	-
A_v	Valve-flow coefficient	m^2	2.1.2
a	Temperature coefficient of linear expansion	$^{o}C^{-1}$	1.1.1.1
a	Coefficient in Antoine's equation	-	1.1.1.3
a	Temperature coefficient of resistance	$^{o}C^{-1}$	1.1.3
B	Thermistor equation coefficient	K	1.1.3
B	Numerical constant	-	-
b	Width	mm	1.1.1.1
b	Temperature coefficient of cubical expansion	$^{o}C^{-1}$	1.1.1.2
b	Coefficient in Antoine's equation	K	1.1.1.3
b	Numerical constant	-	-
C	Specific thermal capacity	$J/kg^{o}C$	5.2
C_v	Valve-flow coefficient	-	2.1.2
c	Controlled variable	^{o}C	5.1
D	Differential operator	s^{-1}	5.2
$D(s)$	Denominator function of s	-	5.3.1
d	Diameter, distance	m	-
E	Modulus of elasticity	kgf/m^2	1.1.1.1
e	2.71828...	-	-
e	Error signal	-	5.1
$F(s)$	Transfer function	-	5.3
Gr	Grashof number	-	Prob. 9
$G(s)$	Transfer function	-	5.7
$H(s)$	Transfer function	-	5.7
h	Hysteresis band width	^{o}C	6.1
h	Heat transfer coefficient	$W/m^{2}{}^{o}C$	5.2
j	$\sqrt{-1}$	-	5.3
K	Gain factor	-	5.4.2
K_p	Proportional gain factor	-	4.3
K_1, K_2, K_3	Valve characteristic coefficients	-	2.1.2
K_v	Valve flow coefficient	-	2.1.2
L	Length	m	1.1.1
L	Distance-velocity lag, pure delay	s	5.9.3

Symbol	Meaning	Units	Section of first appearance
LM	Logarithmic magnitude	dB	5.10.1
$L(s)$	Linear transfer function	-	6.3
$L(f(s))$	Laplace transform of $f(s)$	-	5.3
$L^{-1}(f(s))$	Inverse Laplace transform of $f(s)$	-	5.3
M	Magnitude ratio	-	5.10.1
M	Peak-to-peak amplitude of square wave	-	6.2
m	Output variable of nonlinear element	-	6.2
N	Valve authority	-	2.1.2
$N(s)$	Numerator function of s	-	5.3.1
Nu	Nusselt number	-	Prob. 9
$N(x)$	Describing function magnitude	-	6.3
P	Roots of function of s	s^{-1}	5.3.1
P	Pressure	Pa, mmHg	-
PB	Proportional band	°C, %	4.2
Pr	Prandtl number	-	Prob. 9
Q	Heat-flow rate	W	-
q	Mass-flow rate	kg/s	5.4
$q(t)$	Step response function of linear element	-	6.2
R	Resistance	ohms, °C/W	-
Re	Reynolds number	-	Prob. 9
r	Reference input, set point	-	5.1
r	Frequency ratio	-	5.10.1
s	Laplace operator, $\sigma + j\omega$	s^{-1}	5.3
T	Absolute temperature	K	-
T	Time constant	s	5.2
T_d	Derivative action time	s, min	4.3
T_i	Integral action time	s, min	4.3
$T(\omega)$	Tsypkin function	-	6.4.1
t	Time	s	-
x	Amplitude of a sine wave	-	6.3
χ	Real variable	-	-
Y	Radiation sensitivity factor for thermostat	-	2.7
Z	Roots of a function of s	s^{-1}	5.3.1
α	Diffusion coefficient	kg/s.Pa	1.2.1
β	Imaginary part of a complex number	-	5.10.1
δ	Small real number	-	5.8
ζ	Damping ratio	-	5.6
θ	Temperature	°C	-
π	3.14159...	-	-
ρ	Density	kg/m^3	5.2
σ	Real part of a complex number	-	5.3
ϕ	Phase difference	rad, dig	6.3
ω	Frequency	rad/sec	5.6
∞	Infinity	-	-

CHAPTER 1

SENSORS AND INSTRUMENTATION

1.1 TEMPERATURE SENSING AND INSTRUMENTATION

The primary variable which it is desired to control in building services systems is temperature, usually room air temperature via a heating medium. There exists a very large number of methods of temperature measurement, but we are interested here mainly in those which produce a mechanical or electrical output signal which can be amplified and used to perform a control function, as well as those giving a reading on an indicator.

For convenience, thermometric instruments can broadly be divided into two categories: electric and non-electric. The electric types have the advantage of greater accuracy and ease of transmitting the signal over long distances and interfacing with electronic equipment. Non-electric types are generally cheaper in first cost, are often self-actuating, requiring no external power supply, and they normally produce a mechanical displacement force as the output, which is convenient for use with pneumatic control systems.

1.1.1 Non-electric Methods

One property of solid materials which has been used for many years to measure temperature is thermal expansion. Most materials, solid and fluid, expand on an increase in temperature. This occurs in a very reliable and predictable way since it is a fundamental property of the material, and it requires no external activation to bring it about. Hence, it is a very suitable property for thermometry. Unfortunately, the effect is relatively small for most materials and it is necessary to design the thermometer so that the difference in expansion between two materials is used. In this way the output can be made conveniently large and useful.

1.1.1.1 Solid expansion thermometers. Over small ranges of temperature, the linear expansion of most materials can be described by a simple equation of the form

$$L(\theta) = L_o\,(1 + a\theta) \qquad (1.1)$$

where $L(\theta)$ = length at temperature θ^oC,

L_o = length at 0^oC,

θ = temperature, ^{o}C,

a = temperature coefficient of linear expansion, related to $0^{o}C$.

It is worth specifying the temperature to which the coefficient is related, since temperatures other than $0^{o}C$ have sometimes been used. For temperature measurement we are concerned mainly with expansion of metals. Values of the linear expansion coefficient for a number of metals are given in Table 1.1, from which it can be seen that a is generally in the range $10^{-6}\ ^{o}C^{-1}$ to $10^{-5}\ ^{o}C^{-1}$. If we consider an ordinary steel rod whose length is exactly 1 metre at $0^{o}C$, on heating to $20^{o}C$ the length will become 1.00022 metres, an increase of only about a fifth of a millimetre. This small change would be very difficult and expensive to measure accurately without errors due to the expansion of the measuring equipment itself. It would probably be necessary to make the equipment largely out of Invar, whose coefficient of linear expansion is very small over a moderate range of temperature (up to about $100^{o}C$). This suggests the idea of using the differential expansion between two materials which have differing coefficients. This principle is widely used in bimetal strips and in rod and tube elements. Figure 1.1 shows the arrangement of the two metals for rod and tube elements and for bimetal strips. The rod and tube elements are suitable as sensors for insertion in a vessel or pipe to sense a liquid temperature. The bimetal strips are formed into various shapes before use; straight cantilever, U-shape, spiral or helix, to provide linear, angular or rotational motion for operating electrical contacts or moving a pointer over a dial. Design equations have been developed to predict the deflexion or the force available from a particular size and shape of bimetal.

TABLE 1.1 Linear and cubical coefficients of thermal expansion at normal ambient temperatures

Material	$10^{6}a/^{o}C$ (linear)	Material	$10^{5}b/^{o}C$ (cubical)
Aluminium	23	Ethyl Alcohol	108
Copper	16.7	Mercury	18.1
Brass (^{68}Cu ^{32}Zn)	18-19	n-Pentane	155
36% Nickel Steel (Invar)	0-1.5	Toluene	107
Stainless steel	16	Water	21
Carbon Steel, typical	11	Xylene	112
Glass, Pyrex	3.2		

Quoted, with permission, from Tables of Physical and Chemical Constants, G.W.C. Kaye and T.H. Laby, 14th Ed. Longman, 1973.

The free deflexion d mm of one end of a straight strip which is clamped at the other end is given by

$$d = 1.1\ K\ \ell^{2}\ \Delta\theta/t \qquad (1.2)$$

where ℓ = active length, mm,

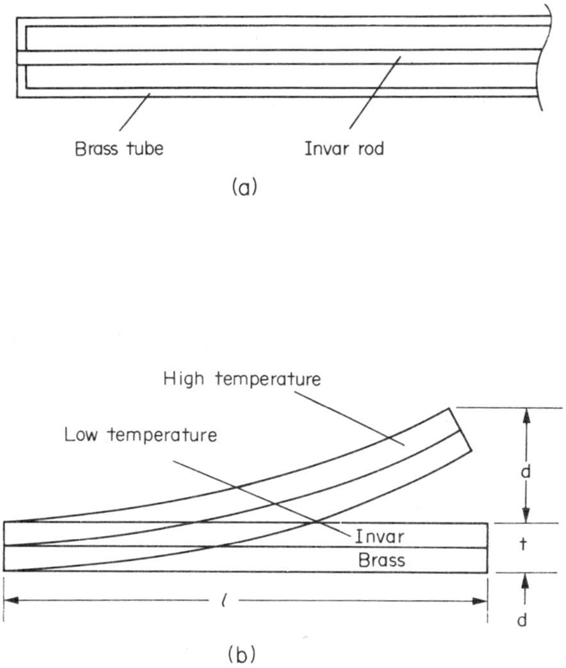

Fig. 1.1. (a) Rod and tube element, (b) bimetallic strip.

t = thickness, mm,

$\Delta\theta$ = change in temperature, ^{o}C,

K = strip deflexion constant, mm/^{o}C.

The strip deflexion constant is defined as 4 times the deflexion of a narrow strip of the bimetal, supported as a beam, per degree C change in temperature over a linear portion of the deflexion-temperature curve. The value of K ranges from about 4×10^{-6} to 20×10^{-6} mm/^{o}C.

The force P kgf, generated by the cantilever which is restrained to prevent movement, is given by

$$P = 1.1\ KEbt^2\ \Delta\theta/4\ell$$

where E = Modulus of Elasticity of the strip, kgf/mm^2,

b = width of strip, mm.

The value of E ranges from about 1.3×10^4 to about 2×10^4 kgf/mm^2.

The useful temperature range is from about -75°C to about 400°C, although some types have a smaller range.

The deflexion produced by rod and tube elements depends only on their length and on the expansion coefficients of the two materials. They can produce large forces, but the small deflexion available makes them unsuitable for use in indicating instruments. Usually a lever arrangement has to be incorporated to magnify the movement and allow operation of contacts or closure of a gas valve.

In recent years an alternative to the bimetal has become available. This is the Shape Memory Effect, discovered in 1958. It is a property of a particular brass alloy, SME brass. A deformation which is applied to the material when it is in a fully martensitic state can be removed progressively by raising the temperature within the range of operation, about 100°C. The deformation reappears as the temperature falls, as if the metal remembered its previous shape. This effect can be used to produce useful deflexions and force similarly to the bimetals, but in a rather more flexible manner.

1.1.1.2 Liquid expansion thermometers. The change in volume of a liquid contained in a constant-volume vessel can be expressed by an equation of the same form as equation (1.1):

$$V(\theta) = V_o (1 + b\theta) \tag{1.4}$$

where $V(\theta)$ = volume at temperature θ°C,

V_o = volume at 0°C

θ = temperature, °C,

b = absolute coefficient of cubical expansion related to 0°C, $^oC^{-1}$.

Values of b are given for a number of liquids in Table 1.1, from which it can be seen that b is generally about 10^{-3} $^oC^{-1}$, except for mercury and water, for which the value is only about one-fifth of this.

If the liquid is contained in a container whose volume is not constant, i.e. one which expands on increase of temperature, then there may be an *apparent* expansion of the liquid. This is expressed by quoting an apparent coefficient of cubical expansion for the liquid. The apparent coefficient of expansion b' for a liquid can be taken as

$$b' = b - 3.a \tag{1.5}$$

where b = absolute coefficient of cubical expansion of the liquid

and a = coefficient of linear expansion of the material of the container.

The commonest form of liquid expansion thermometer and the one with the longest history is the mercury-in-glass thermometer. The transparency of the glass allows the thread of mercury to be seen easily and the temperature read against a scale scratched or etched on the glass surface. Mercury is an excellent thermometric liquid in that its coefficient of expansion is very constant with temperature, although rather small, there is a wide temperature range between the freezing and boiling points (-39°C and 356°C respectively). It is possible to extend the lower end of the range down to about -55°C by using an amalgam of mercury with thallium. The upper limit can be raised to about 450°C by filling the space above the mercury thread with nitrogen, which is compressed as the mercury thread rises and defers boiling to temperatures well above the normal boiling point. The upper temperature limit is fixed by the softening of the glass.

The small size of the coefficient of expansion of mercury necessitates the use of rather fine-bore capillary tubes in mercury-in-glass thermometers. This makes the thread of mercury very fine and because of the mirror-like colourless surface of the mercury it can be difficult to see. It is possible to improve visibility of the mercury thread by incorporating a red glass ribbon in the stem so that the mercury appears red by reflection. There is a need, however, for a cheaper thermometric liquid for non-critical applications and this can be met by the use of alcohol with a dye to make it easily visible. The large coefficient of expansion allows wider bore capillaries to be employed. For low-temperature work, liquid-in-glass thermometers have been used with toluene or pentane, allowing use down to -80 and -200°C respectively.

Dial Thermometers. The liquid-in-glass thermometer is useful as a direct indicating instrument, being fairly cheap and accurate and needing no external source of power. However, there exist many industrial situations where it is necessary to be able to read the temperature at a point distant from the sensing point. It is desirable, for example, to have an indication on a control panel of the temperatures at a number of points in the plant, which may be widely separated and distant from the plant room. In response to this requirement a type of expansion thermometer has been developed where the expanding liquid is contained within a metallic casing and completely fills it. There is no meniscus as with liquid-in-glass thermometers. The temperature sensor is a cylindrical bulb which is filled with the liquid. Changes in the volume of the liquid due to temperature change are transmitted by a flexible metal capillary tube to an indicating device which is usually a Bourdon tube. The capillary tube can be up to about 50 m long, as in Fig. 1.2. The bore is about 0.25 to 0.5 mm. The filling liquid may be mercury,

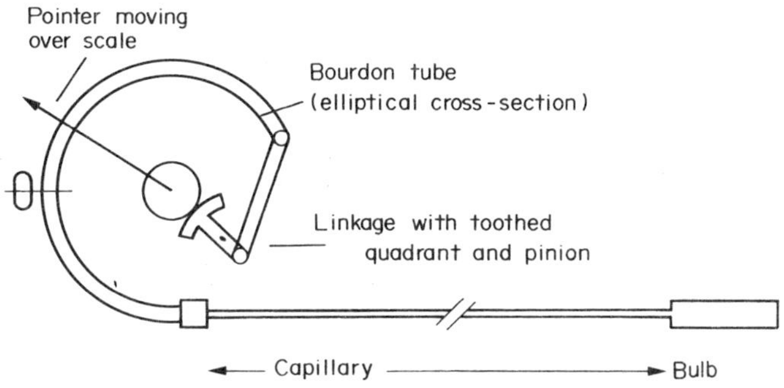

Fig. 1.2. Dial thermometer with Bourdon tube and indicator.

in which case the bulb and capillary must not be made of a copper alloy so that amalgamation will not occur: stainless steel would normally be used. For a cheaper system the filling liquid could be xylene or alcohol, each of which has a coefficient of cubical expansion about 6 times as great as that of mercury. These systems are filled under pressure and can be used up to temperatures well above the normal atmospheric boiling point of the filling liquid.

They are subject to reading errors due to two effects: head errors caused by the relative levels of the dial and the bulb being different on installation from what they were on calibration, and temperature errors caused by a sensitivity to changes in ambient temperature at the capillary and at the dial, not only at the bulb. The head error can be removed by a manual adjustment at the time of installation. The temperature errors are reduced by inserting links containing Invar wires in the capillary and by including an automatic compensator in the dial mechanism, using a bimetallic strip. Alternatively both effects can be compensated for automatically by using a second capillary tube adjacent to the main tube and connected to a second Bourdon tube, arranged in opposition to the first. The second capillary tube has no bulb and its only function is to compensate for errors, as indicated in Fig. 1.3. This system is of course more expensive than the simple uncompensated ones.

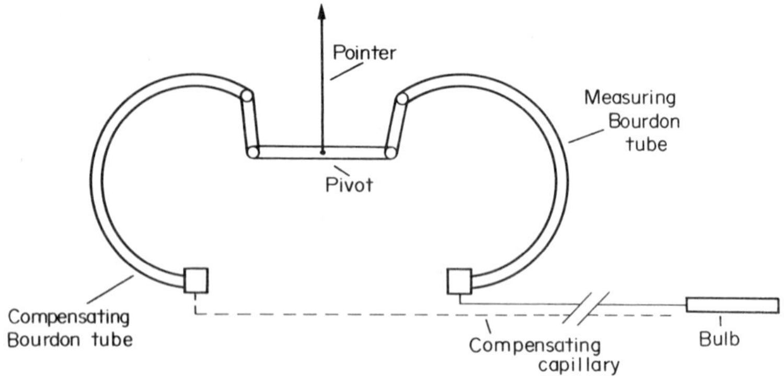

Fig. 1.3. Double capillary and Bourdon tube system for compensation. Note that the compensating capillary is not connected to the bulb.

These systems have tended to be replaced in recent years by electrical ones, particularly those using thermistors, where a single indicator can easily be switched among a large number of sensing points.

1.1.1.3 Vapour pressure thermometers. Consider the system shown in Fig. 1.4, where the bulb contains the interface between the liquid form and the saturated vapour of a volatile substance. All air is excluded from the system, which contains only the volatile substance in liquid or vapour form. The pressure

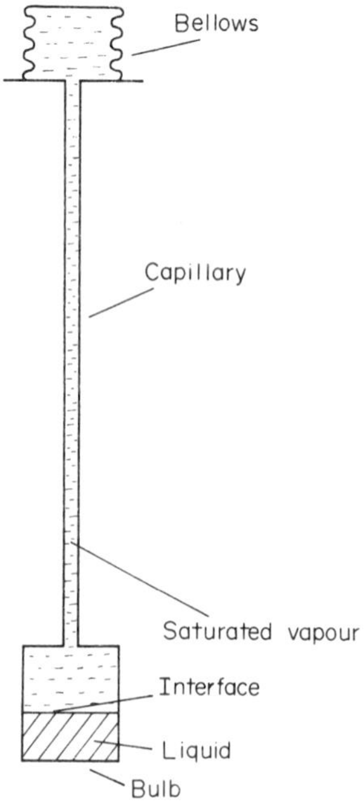

Fig. 1.4. A vapour-pressure system.

existing within the system will be the saturated vapour pressure (SVP) of the substance at the temperature of the interface. If the temperature rises, some of the liquid will boil off until equilibrium is reached, and similarly if the temperature falls some of the vapour will condense, reducing the vapour pressure until an equilibrium is again reached at a lower SVP. The SVP/temperature relationship for a number of substances is shown in Fig. 1.5 and is expressed empirically by Antoine's equation:

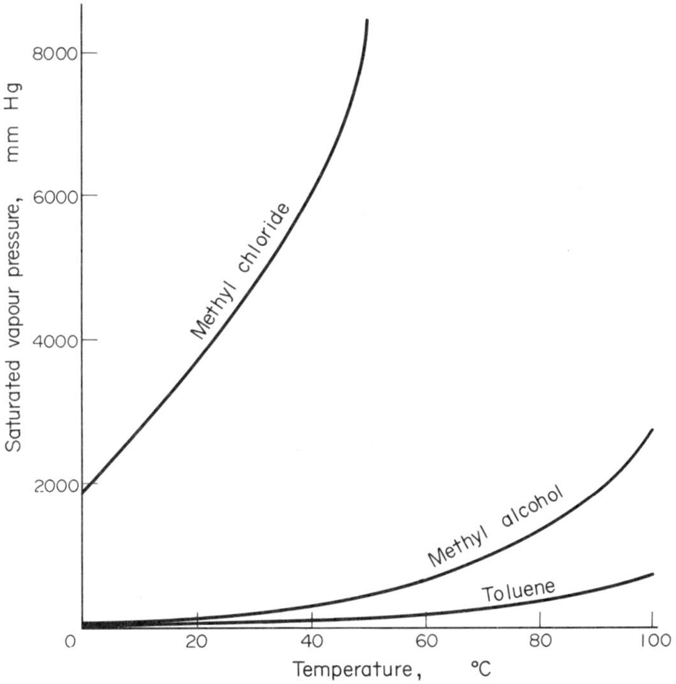

Fig. 1.5. SVP-temperature curves.

$$\log_{10} P = a - b/T \tag{1.6}$$

where P = SVP, mm of mercury,

T = absolute temperture, ^{o}K

and a, b are constants depending on the substance:

	a	b
Methyl alcohol	8.80	2001
Methyl chloride	7.48	1148
Toluene	8.33	2047

The calibration of this type of thermometer is therefore dependent only on the properties of the substance with which it is filled and not, for example, on the volume of the bulb and capillary as with a liquid expansion system. The pressure inside is relatively low, so the thickness of the metal used can be kept small. The instruments do not need to be calibrated individually. These factors mean that these instruments can be produced much more cheaply than liquid-filled systems. The dial indicators fitted to domestic solid fuel boilers are almost always of this type, giving a cheap reliable indication of water temperature. The scales on these instruments are not uniform, the divisions being more widely separated at higher temperature.

This principle is also used in some room thermostats, the volatile substance being contained in a capsule or bellows which responds to the surrounding air temperature. As the temperature changes, the pressure exerted by the substance changes, producing a physical displacement of the bellows and operating a set of electrical contacts or a microswitch. Those vapour-pressure thermometers where a bulb and dial are connected by a capillary tube have two modes of working, depending on whether the indicator and capillary are at a higher or lower temperature than the bulb. In the former case the vapour is at a higher temperature than its dew-point and does not condense. In the latter case the vapour will condense and eventually the whole of the cooler part will be filled with liquid. There may then be errors due to the pressure of the column of liquid in the capillary, and possibly due to the interface between liquid and vapour not being in the bulb. This type of instrument is best suited to measuring temperatures which are either always above ambient or always below ambient: this avoids the so-called "cross-ambient effect". Systems which are not subject to the cross-ambient effect errors are also available. The volatile liquid and its vapour are confined within the bulb, and the pressure changes are transmitted to the bellows by an inert liquid which fills the bellows, capillary and part of the bulb.

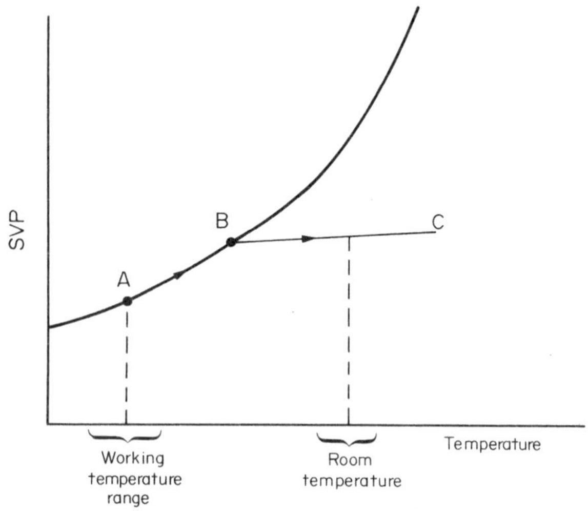

Fig. 1.6. Illustrating the fade-out effect.

Where these instruments are used to measure or control refrigeration systems, the working temperature is well below ambient and the bellows or Bourdon tube has to be relatively sensitive to respond to the lower pressure available from that part of the SVP/temperature curve. When the refrigeration plant is switched off, its temperature will rise to ambient and the pressure in the system will rise accordingly. To avoid damage to the bellows caused by the high pressure, the "fade-out" effect is produced. The system is filled only with a small quantity of the volatile substance, sufficient to provide some liquid at low temperatures, as in Fig. 1.6, point A. As the temperature rises, more and more of the liquid evaporates until at point B the system is entirely filled with saturated vapour and no liquid is left. When the temperature rises further, any increase in pressure is due only to expansion of the vapour, and follows the line BC instead of continuing up the SVP curve. The pressure-sensitive element is thus protected from damage by excessive pressure.

1.1.2 Electrical Thermometers-Thermocouples

The methods described above involving thermal expansion are suitable for temperatures up to a maximum of about 500°C. Where it is required to measure temperatures up to about 2000°C it is necessary to employ electrical methods; in particular the thermoelectric method. A thermocouple is a closed circuit of two metals. When the two junctions between the metals are at different temperatures, a thermoelectric e.m.f. (electromotive force) is set up between the junctions, and causes a current to flow round the circuit. The e.m.f.s available from most thermocouple materials are not large and generally the value of e.m.f. is a rather complicated and irregular function of the junction temperatures. While it is possible to derive theoretical expressions for the e.m.f. as a function of temperature, they are not well followed in practice. Usually it is necessary to consult tables of e.m.f. versus temperature for the various types of thermocouple in use, such as those published by the British Standards Institution (BSS 4937, parts 1 to 7, 1973). For noble metal thermocouples, made from platinum and its alloys, the e.m.f. obtainable is only about 6 μV/^{o}C and may be only one-twentieth of this for the highest-temperature types. For base-metal thermocouples, which do not include platinum, the voltage output is much greater, generally about 40 μV/^{o}C, but the upper temperature limit is lower. The most commonly used types and their characteristics are summarised in Table 1.2.

The thermal e.m.f. is dependent on the temperatures of the two junctions of the thermocouple. If it is required to measure the temperature of one of the junctions, the measuring junction or "hot" junction, then the temperature of the other, the reference or "cold" junction, must be known or must be fixed in some way. There are basically three possible approaches for dealing with this requirement.

1. The cold junction may be maintained at a fixed point whose temperature is defined, e.g. 0°C in melting ice or more conveniently in one of the automatic ice-point cells which are commercially available. This method is the only acceptable one for measurements of high accuracy.

2. The cold-junction temperature may be allowed to "float" at ambient temperature. This temperature is measured with a subsidiary thermometer, usually mercury-in-glass, and a correction is applied.

3. The cold-junction temperature is allowed to float in a similar way to the previous method, but an automatic compensation circuit is included in the thermocouple circuit. The compensation circuit automatically injects a voltage equivalent to the difference between the ambient temperature and 0°C. As the ambient temperature varies, the voltage produced by the compensation circuit varies in sympathy: the overall output e.m.f. from the thermocouple is always that which would have been obtained if the cold junction were really at 0°C. This method gives compensation for changes in ambient temperature over a moderate range. It is suitable for direct-reading instruments of moderate accuracy. A typical circuit to give automatic cold-junction compensation is shown in Fig. 1.7, where a thermistor is used as the sensing element for the compensating circuit. A Wheatstone bridge is arranged to produce the compensating voltage.

Thermocouple properties are sometimes described by the so-called Three Laws of Thermoelectricity.

1. Law of the Homogeneous Circuit: an electric current cannot be maintained in a

TABLE 1.2 Thermocouple types and ranges

Type letter	BS 4937 part:	Positive metal	Negative metal	e.m.f. for 100°C (microvolts)	e.m.f. for 1000°C (microvolts)	Temperature range (°C)
S	1	Platinum +10% Rhodium	Platinum	645	9,585	-50 to 1760
R	2	Platinum +13% Rhodium	Platinum	647	10,503	0 to 1830
J	3	Iron	Copper +40% Nickel	5268	57,942	-210 to 1200
K	4	Chromium + Nickel	Aluminium + Nickel	4095	41,269	-270 to 1370
T	5	Copper	Copper +40% Nickel	4277	20,869 (at 400°C)	-170 to 400
E	6	Chromium + Nickel	Copper +40% Nickel	6317	76,358	-270 to 1000
B	7	Platinum +30% Rhodium	Platinum +6% Rhodium	33	4,833	0 to 1830

Extracts from BS 4937, Parts 1 to 7, are reproduced by permission of BSI,2 Park St, London W1A 2BS, from whom complete copies can be obtained. Proprietary names for alloys: chromium+nickel=chromel, aluminium+nickel=alumel, copper+40% nickel=constantan.

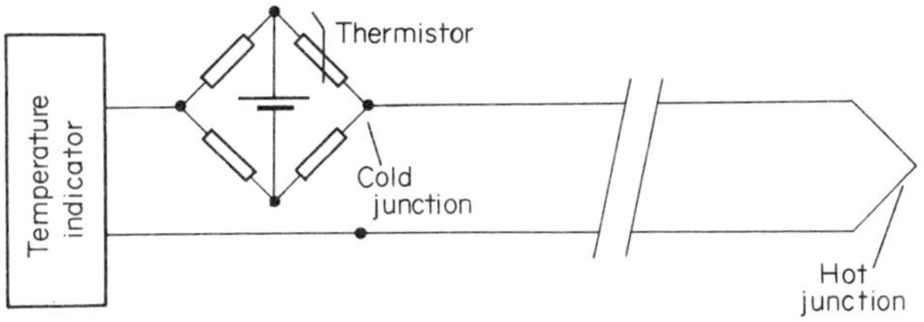

Fig. 1.7. Circuit for automatic compensation for ambient temperature changes.

single homogeneous circuit by the application of heat alone. The operative word here is "homogeneous". This law is sometimes used to test a sample of wire for homogeneity. A sensitive galvanometer is connected to the two ends of the wire and a flame is traversed slowly along the length of the wire. The magnitude of the galvanometer deflections gives a measure of the variation in composition or purity along the wire, and portions of high impurity level can be identified and avoided. This technique is useful for work of high precision.

2. Law of Intermediate Metals: if a thermocouple consists of two metals A and B, and produces an e.m.f. e when the junctions are at fixed temperatures, then e is not affected by the introduction of a third metal C, if the ends of C are at equal temperatures. This is illustrated in Fig. 1.8, which shows that this

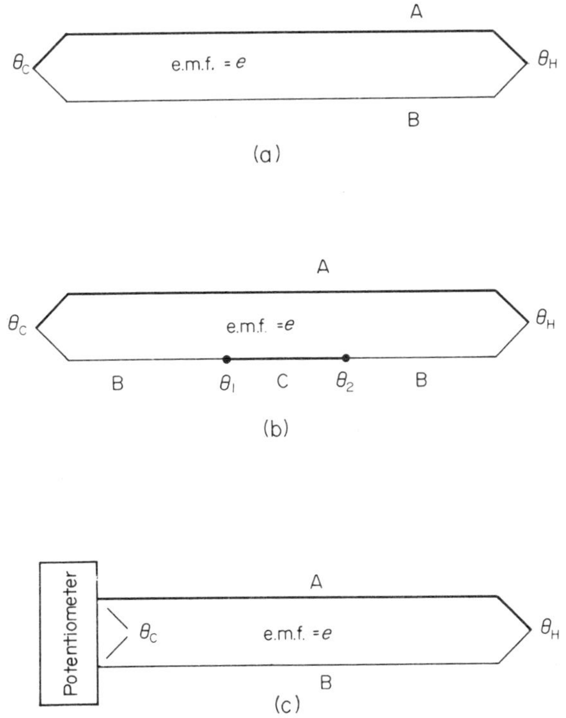

Fig. 1.8. (a) Simple thermocouple producing e.m.f. e. (b) Same e.m.f. produced if $\theta_1=\theta_2$. (c) Same e.m.f. produced if the terminals of the potentiometer are both at θc.

law allows a measuring instrument to be connected in to the thermocouple without influencing the e.m.f., so long as its terminals are both at the same temperature.

3. Law of Intermediate Temperatures: the e.m.f. of a thermocouple whose junctions are at temperatures θ_1 and θ_3 is equal to the algebraic sum of the e.m.f.s of two similar thermocouples whose junctions are at temperatures θ_1 and θ_2 and θ_3 respectively. This law can be regarded as a statement of the fact that the thermocouple characteristic is not uniform, but is an irregular function of temperature. Doubling the temperature difference does not generally produce an exact doubling of the e.m.f. output.

The hot junction temperature of the thermocouple may be indicated in two alternative ways: the current round the circuit may be allowed to flow through a galvanometer or milliameter whose deflection gives a reading of temperature, or the e.m.f. may be measured by a potentiometer or high-impedance d.c. millivoltmeter. In the first of these methods the indicator is responding to the current in the circuit, which is dependent on the circuit resistance as well as the e.m.f. This means that changes in the circuit resistance will alter the calibration and introduce errors in the reading. On installation, the calibration has to be set with regard to the particular length, type and gauge of wire used. In use, because the coil of the galvanometer is wound of fine copper wire whose resistance is dependent on temperature, changes in ambient temperature will produce spurious changes in reading. This type of error can be reduced by the inclusion in the circuit of a "swamp" resistor. This is a resistor made of a metal, such as manganin or constantan, whose resistance value is not dependent on temperature. The variations in galvanometer resistance are "swamped" by the large constant value of the swamp resistor. The error cannot be removed completely but it can be made negligibly small.

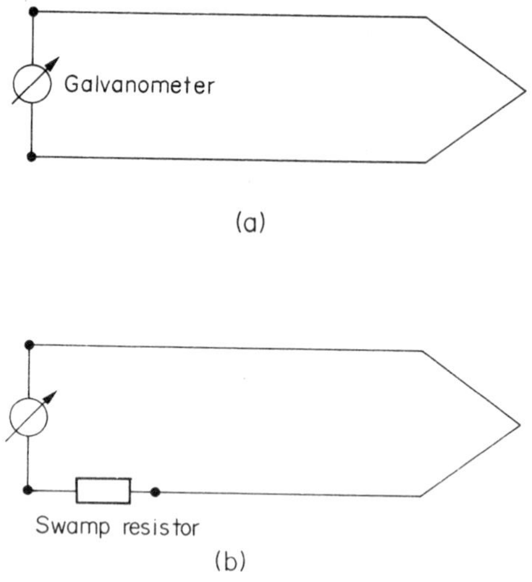

Fig. 1.9. (a) Thermocouple circuit. (b) With swamp resistor.

Where the e.m.f. is measured by a high input-impedance circuit such as a potentiometer or digital voltmeter, there is almost zero current flow, so the circuit resistance does not affect the readings. This method is preferred for measurements of high accuracy. There are a number of instruments available which automatically convert the thermocouple e.m.f. into temperature and display the result in either digital or analogue form.

Because of the small output voltages available from thermocouples, they are not normally used for sensors in automatic control circuits. They are more usually employed for indication, particularly of high temperatures.

1.1.3 Electrical Resistance Thermometry

Many materials undergo a change in their electrical resistance with temperature. This is a reliable and reproducible effect and is convenient both for very precise measurement and for control of temperature. For moderate ranges of temperature the variation of resistance can be expressed by an equation of similar form to equation (1.1).

$$R(\theta) = R_o\,(1 + a\theta) \qquad (1.7)$$

where $R(\theta)$ = resistance at temperature θ^oC,

R_o = " " 0^oC,

θ = temperature, ^oC,

a = temperature coefficient of resistance, related to 0^oC, $^oC^{-1}$.

The values of the coefficients of resistance for a number of materials are listed in Table 1.3. The material whose resistance behaviour as a function of temperature has been studied most intensively is platinum. The platinum resistance thermometer is the standard reference thermometer over the range -185^oC to 660^oC. Nickel is a much cheaper material with a resistance coefficient about 50% greater than that of

TABLE 1.3 Temperature coefficients of electrical resistance for a number of materials

Material	Temperature coefficient of electrical resistance ($^oC^{-1}$)	Upper temperature limit (^oC)
Platinum	3.85×10^{-3}	660
Nickel	6×10^{-3} approx.	300
Copper	4.0×10^{-3}	120
Balco (40 Ni, 60 Fe)	4.1×10^{-3}	250
Constantan (40 Ni, 60 Cu)	0.02×10^{-3}	500
Manganin (18 Mn, 4 Ni, 78 Cu)	0.04×10^{-3}	500
NTC Thermistor (semiconductor)	-40×10^{-3} approx.	1000

(Constantan is also known by the proprietary names; Eureka & Advance.)

platinum, but the coefficient is very variable with temperature. The resistance of copper changes quite markedly with temperature, and it will be seen that the variation of resistance with temperature of copper connecting leads has to be compensated for. Balco is the proprietary name of an alloy which has a uniform and fairly large variation of resistance with temperature. Occasionally it is necessary to have materials whose electrical resistance is invariant with temperature, and constantan and manganin are alloys whose coefficients of resistance are two orders of magnitude smaller than those of most metals. In recent years the thermistor, consisting of a semiconductor material, has increased enormously in number and range of applications. Early objections to the lack of stability and small temperature range of thermistors have now been overcome by improvements in the manufacturing process. In contrast to metallic materials, NTC thermistors exhibit a reducing resistance as temperature rises. They have a negative temperature coefficient, hence NTC. Alternatively, it is possible to obtain semiconductors whose resistance increases with temperature (hence PTC). These are not normally used for temperature measurement, however, because their resistance "jumps" by up to four orders of magnitude over a very narrow range of temperature, and is nearly constant outside that range, as shown in Fig. 1.10. PTC thermistors can be used

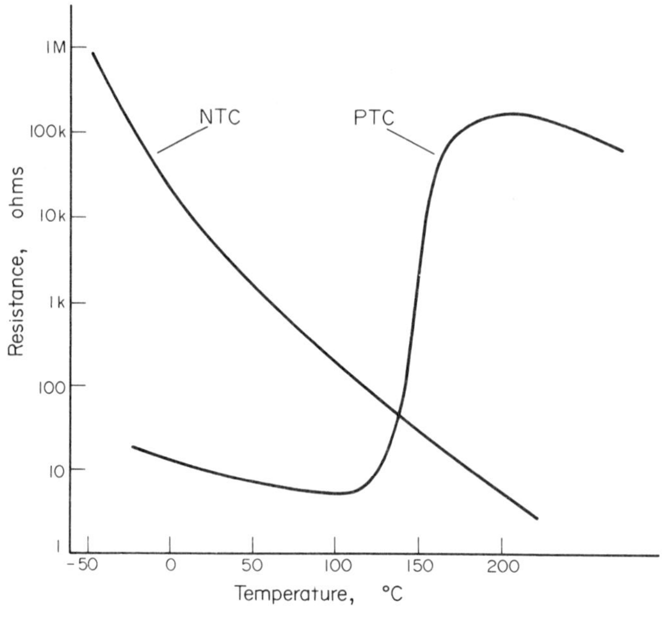

Fig. 1.10. Typical resistance-temperature characteristics of NTC and PTC thermistors. Note the logarithmic scale of resistance.

for control purposes, one popular application being as a temperature-limiting device in the windings of electric motors. The PTC thermistor is connected in series with the copper windings, and is chosen to have a negligibly small resistance at ordinary load conditions. If the motor is overloaded and the temperature rises above the switching point, the resistance rises very rapidly, cutting off the current until the motor has cooled.

The variation with temperature of the resistance of NTC thermistors is usually described by an exponential type of equation of the following form:

$$R(\theta) = A \cdot \exp(B/T) \tag{1.8}$$

where $R(\theta)$ = resistance, ohms, at temperature $\theta^{o}C$,

T = absolute temperature, ^{o}K,

A = a constant dependent on the size and shape of the thermistor element, ohms,

B = the material constant of the thermistor element, ^{o}K.

The factor A obviously corresponds to the thermistor resistance at infinite temperature. The factor B is not truly constant for a given thermistor; it varies slightly with temperature, but generally has a value between 3000 and 4000°K.

Taking the natural logarithm of equation (1.8) gives

$$\ell n\ R = \cdot \ell n\ A + B/T \tag{1.9}$$

which can be regarded as the equation of a straight line of slope B and intercept ℓn A. The values of A and B can thus be determined experimentally by measuring the resistance at various temperatures and plotting ℓn R against 1/T. The intercept at 1/T = 0 gives the value of ln A. The slope of the line gives the value of B.

Thermistors are formed from a sintered ceramic made from a mixture of oxides of the elements Mn, Ni, Co, Cu and Fe. The resistance can be chosen to be between 10^{0} and 10^{6} ohms at room temperature, by varying the size, shape and composition of the elements. Coefficients of resistance from about -20×10^{-3} to about $-65 \times 10^{-3}\ ^{o}C^{-1}$ can be obtained. These devices are available in bead, disc and rod form, varying in size from beads of 0.2 mm diameter to rods up to 40 mm long. They may be encapsulated in solid glass or housed in glass envelopes or transistor cans. They may be self-heated by current passing directly through them or indirectly heated by a wire winding. In addition to their use for temperature measurement, they can be used for air-flow measurement in anemometers, for liquid flow measurement and for sensing of the level of liquid in a vessel. Types are available for use up to 1000°C.

For some applications the very non-uniform resistance-temperature characteristic is a disadvantage, and it is required to obtain a more linear variation of resistance with temperature. Figure 1.11 shows how this can be done most simply by connecting a fixed resistor in parallel with the thermistor. The value of the fixed resistor is chosen to be equal to the resistance of the thermistor at the centre of the temperature range where linearization is required, θ_1. At high temperatures the resistance of the thermistor drops and the resistance of the combination approaches the value of the thermistor. At low temperatures the resistance of the thermistor increases and the resistance of the parallel combination approaches the value of the fixed resistor. In between these limits the variation is very uniform, but some sensitivity has been sacrificed to obtain this linearity. Better linearization can be obtained by using arrangements with two thermistors and a number of fixed resistors in series and parallel.

Figure 1.12 shows the basic Wheatstone bridge circuit, which can be used with a metallic temperature sensor or a thermistor for temperature measurement and control.

If we apply Ohm's law to the branch currents i_1 and i_2 we have (ignoring R_5):

$$i_1 = V_s\ (R_1 + R_2), \tag{1.10}$$

$$i_2 = V_s\ (R_3 + R_4)$$

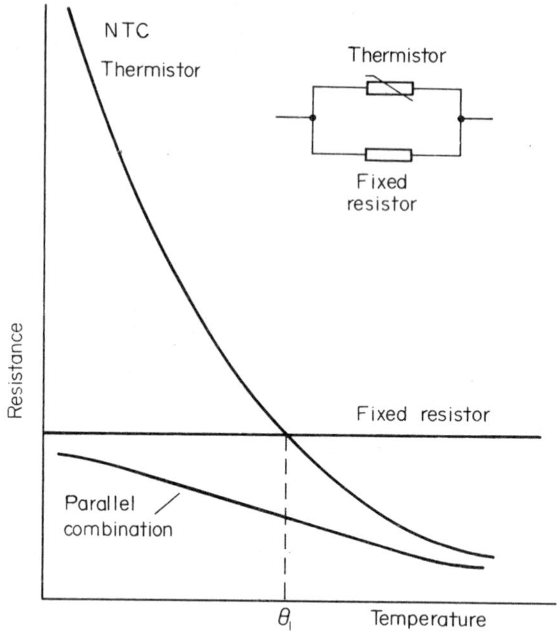

Fig. 1.11. Linearization of a thermistor by a fixed resistor in parallel

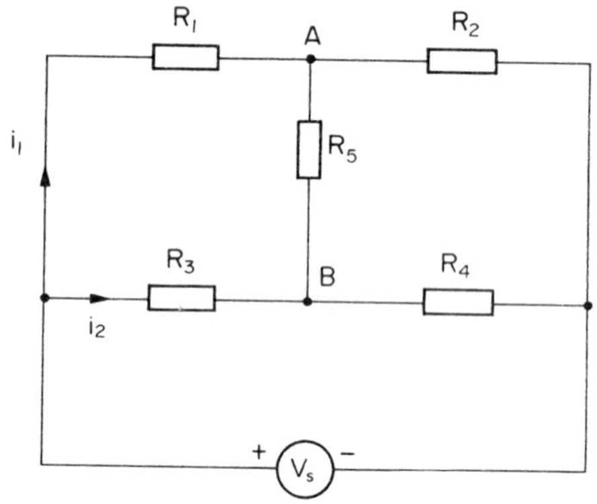

Fig. 1.12. Basic Wheatstone bridge circuit.

where V_s is the supply voltage.

The voltages V_1 and V_3 across resistors R_1 and R_3 are then given by

$$V_1 = i_1 \cdot R_1 = V_s \cdot R_1/(R_1 + R_2), \tag{1.11}$$

$$V_3 = i_2 \cdot R_3 = V_s \cdot R_3/(R_3 + R_4).$$

The bridge output can be regarded as the voltage appearing between the points A and B, the junctions of the arms of the bridge. This is obviously the difference between the voltages V_1 and V_3:

$$V_o = V_1 - V_3 = V_s(R_1/(R_1 + R_2) - R_3/(R_3 + R_4)). \qquad (1.12)$$

When V_o is zero the bridge is said to be balanced. This can occur only if

$$R_1/(R_1 + R_3) = R_3/(R_3 + R_4) \qquad (1.13)$$

and simplifying equation (1.13) gives the three alternative forms for the relationship which the four arms of the bridge must satisfy for balance:

$$R_1 R_4 = R_2 R_3,$$

$$\text{or} \quad R_1/R_2 = R_3/R_4 \qquad (1.14)$$

$$\text{or} \quad R_1/R_3 = R_2/R_4.$$

This simple analysis assumes no direct connection between points A and B through which a current could flow, such as a galvanometer or the input impedance of an amplifier. If such a component of resistance R_5 is connected between points A and B in Fig. 1.12 then the output voltage V_o, expressed as a fraction of the supply voltage V_s, becomes:

$$V_o/V_s = \frac{R_5 (R_1 R_4 - R_2 R_3)}{R_2 X + R_1 Y} \qquad (1.15)$$

where $X = R_1 R_3 + R_1 R_4 + R_3 R_4 + R_3 R_5 + R_4 R_5$

and $Y = R_3 R_4 + R_3 R_5 + R_4 R_5$

by solution of the three loop equations, and elimination of i_1 and i_2.

It often happens that the value of R_5 is large compared with the values of R_1 to R_4. In this case equation (1.15) can be simplified to give the following expression for the output voltage:

$$V_o/V_s = (R_1 R_4 - R_2 R_3)/(R_1 + R_2)(R_3 + R_4) \qquad (1.16)$$

It can be shown that the bridge sensitivity (i.e. the change in V_o/V_s for a given small change in one of the resistors) is greatest when the four resistances R_1 to R_4 are all equal. Hence, if the value of the sensor R_1 is usually about 100 ohms, it is generally best to choose R_3 and R_4 to be 100 ohms also as well as R_2. The peak of sensitivity is a broad one, however, and the output will be reduced by only about 10% if R_2 and R_4 are both chosen to be, say, 200 ohms, with R_1 and R_3 at 100 ohms.

For simple applications of the Wheatstone bridge to temperature measurement, there is generally a single sensor whose resistance changes as the temperature changes, the other three resistances remaining constant. If we take R_1 in Fig. 1.12 as this sensor and differentiate equation (1.16) to obtain the slope of the output voltage, we have:

$$\frac{d(V_o/V_s)}{dR_1} = \frac{R_2}{(R_1 + R_2)^2} \qquad (1.17)$$

and this derivative approaches 0.25 as the value of R_1 approaches the value of R_2. This means that, for each unit of supply voltage V_s, the initial rate of change of output voltage V_o will be 0.25 volt per ohm change in R_1. This initial rate of change reduces quite rapidly, becoming about 0.1 when R_1 is equal to $2R_2$. The bridge sensitivity can sometimes be increased by using two similar sensors connected in diagonally opposite arms of the bridge (R_1 and R_4 or R_2 and R_3) and both sensing the same temperature. In applications such as the measurement of a temperature difference, two sensors can be used in adjacent arms of the bridge (R_1 and R_2 or R_3 and R_4) or the sensitivity can again be increased by using four sensors, connected one in each arm of the bridge. The high-temperature side would be sensed by sensors in the positions of R_1 and R_4, the lower temperature by sensors in the positions of R_2 and R_3. This principle is also employed in other applications, such as strain gauges, where two or four sensors can be placed close together to eliminate errors due to temperature variations.

There are two alternative modes of operation of the Wheatstone bridge for temperature measurement: the out-of-balance mode, in which no adjustment is made to the bridge resistances and current is allowed to flow through a galvanometer to give a deflection, which is read off a scale of temperature; and the null-balance mode, in which the ratio of resistances R_3 and R_4 is adjusted to bring the bridge always into balance. The galvanometer or millivoltmeter is used simply as a null detector. The temperature setting is denoted by the adjustment required of R_3 and R_4 to give a null balance. Obviously the out-of-balance mode is simpler and quicker in use since no adjustment is necessary and readings can easily be taken by unskilled personnel. A high-impedance digital voltmeter can be used as an indicator, and the display can be scaled to read directly in temperature units. The calibration is dependent on the supply voltage being constant. The null-balance mode requires some adjustment to be made to the bridge, so it needs rather more skill to operate, but its calibration depends only on the constancy of resistance values, and not on the supply voltage being maintained constant. Hence it is capable of far greater precision than is the out-of-balance mode.

In the out-of-balance mode, the variation of output voltage is not a straight-line function of detector resistance. Figure 1.13 shows how V_o/V_s varies with the

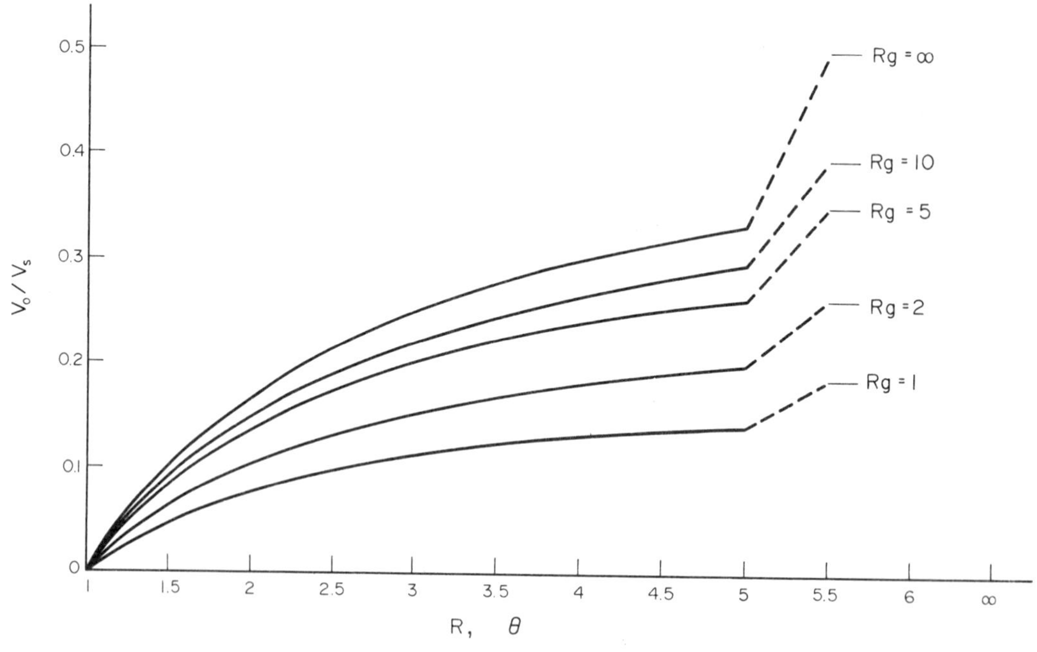

Fig. 1.13. Variation of bridge output ratio V_o/V_s with $R(\theta)$ for various values of galvanometer resistance Rg. The other bridge resistances R_2, R_3 and R_4 are taken as 1.0 throughout.

value of R_1 for various values of R_5. The other three resistances R_2, R_3 and R_4 are assumed to be fixed at unity.

Figure 1.14 illustrates a Wheatstone bridge arrangement for use in the null-balance mode, where the detector R(θ) may be some distance from the bridge. The resistance

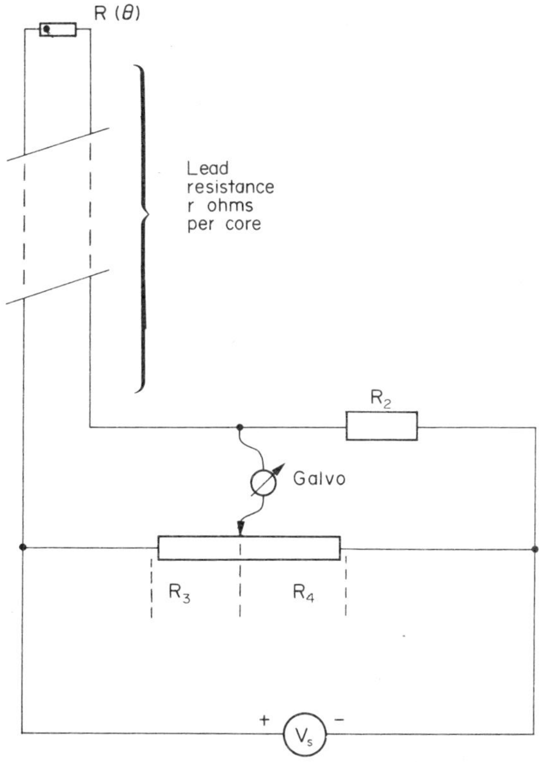

Fig. 1.14. Two-wire bridge used in the null-balance mode.

of each core of the cable connecting the detector with the rest of the bridge circuit may not be negligible if the run of cable is long. If this core resistance is denoted by r, we can write the expression for balance as

$$(R(\theta) + 2r)/R_3 = R_2/R_4 \qquad (1.18)$$

The resistance 2r appears as part of one of the bridge resistances and its value will affect the balance condition. There are two ways in which this lead resistance affects the calibration. On installation, the calibration will be affected by the length of cable used, and it will be necessary to make some adjustment for the actual resistance of that length of cable. In use, when the temperatures to which the cable is exposed change, the resistance of the copper cores will change and cause errors in reading. This drawback can be avoided by the use of the three-wire bridge arrangement, as shown in Fig. 1.15. The expression for balance given in equation (1.14) is now modified as follows

$$(R(\theta) + r)/(R_3 + r) = R_2/R_4. \qquad (1.19)$$

The lead resistance now adds in to both the numerator and denominator of the left-

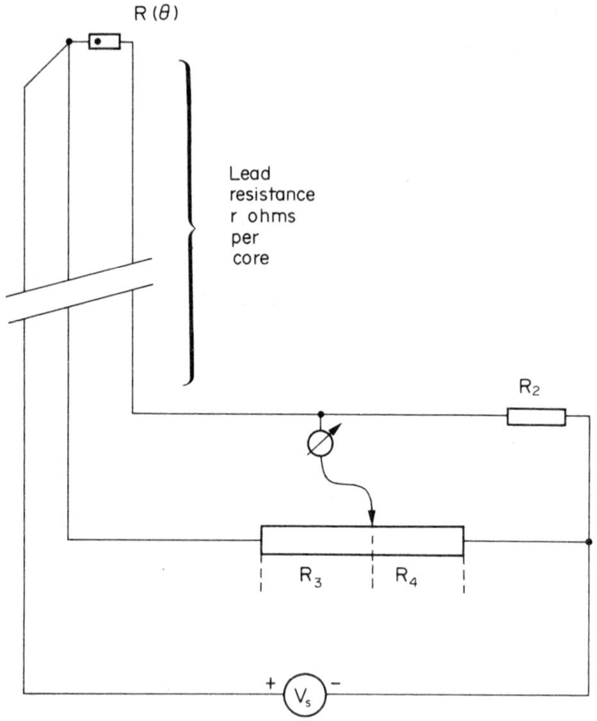

Fig. 1.15. Three-wire bridge used in the null-balance mode, with automatic compensation for lead resistance variations.

hand side. One core resistance is in series with the power supply and does not affect the balance. It can be seen that, although the lead resistance does not actually cancel out from equation (1.19), it now affects both numerator and denominator equally. Since the value of r is always small compared with the values of $R(\theta)$ and R_2, the effect is that the left-hand side of equation (1.19) is very nearly independent of changes in the value of r. This form of bridge is the one almost invariably used for practical measurement and control purposes.

It is sometimes necessary to have a circuit which responds, not to a single temperature, but to the difference in temperature between two points. Figure 1.16 illustrates a bridge which has a differential mode of measurement, the polarity and magnitude of the output voltage being dependent on the sign and magnitude of the difference $R(\theta_1) - R(\theta_2)$. If the two temperature detectors have matched resistance/temperature characteristics then the output voltage will correspond to the difference in temperature. The bridge sensitivity can be doubled by replacing the fixed resistors R_3 and R_4 by two more matched temperature detectors with similar characteristics to those of $R(\theta_1)$ and $R(\theta_2)$. The detector $R(\theta_4)$ must be placed so as to sense the same temperature as $R(\theta_1)$, and $R(\theta_3)$ the same temperature as $R(\theta_2)$. The bridge is then used in the out-of-balance mode and its output voltage will be linearly proportional to temperature difference for small changes of temperature. The output is also proportional to the supply voltage, and this fact can be made use of in a method for monitoring the heat output of a hot-water boiler or of a hot-water heating system. The heat emission is proportional to the product of water-flow rate and temperature difference between flow and return water. A differential bridge is used, with temperature sensors in the flow and

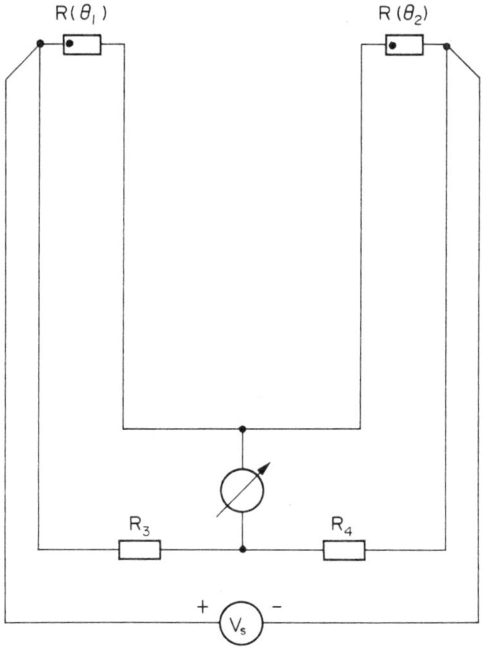

Fig. 1.16. Wheatstone bridge used for measuring temperature difference.

return mains. The supply voltage to the bridge is derived from a unit which monitors the rate of flow through the boiler. The bridge output voltage is therefore proportional to the product of flow rate and temperature difference and hence is a measure of the heat-emission rate.

1.2 HUMIDITY

A simple space-heating system has the power to influence only the dry-bulb temperature in a space, and this is the main requirement for human comfort control, since air temperature is the most important variable which influences comfort. For any given dry-bulb air temperature, the air humidity can vary from perfectly dry to completely saturated, and human comfort is affected particularly at these extremes of humidity. For the best quality of comfort air-conditioning, it is necessary to ensure that the humidity of the air lies within a reasonable band. For certain process requirements, such as in the printing industry, the humidity must be maintained within close limits for operational reasons. There is thus a requirement for a variety of humidity sensors, able to produce mechanical or electrical output signals in the same way as temperature sensors, to give indication and control of humidity.

It should be mentioned that humidity is not a single variable like temperature, but has a number of aspects. Figure 1.17 shows a sketch of a psychrometric chart illustrating the various properties which, in combination with dry-bulb temperature, define the condition of air. The line AA^1 is a contour of constant relative humidity (%) or percentage saturation; the line BB^1 is a line of constant moisture content (kg water/kg dry air), or constant dew point (^{o}C), or constant vapour

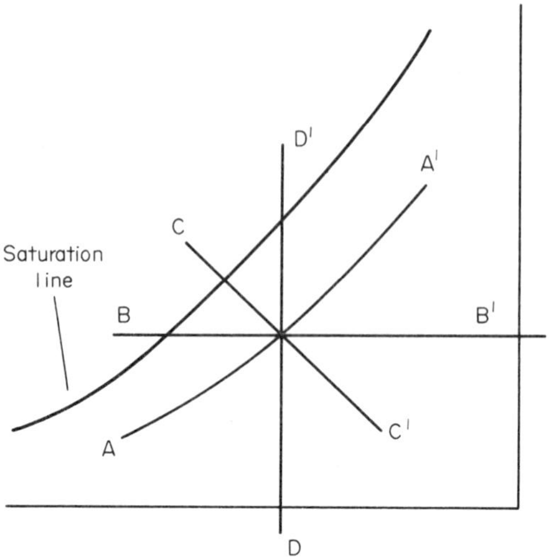

Fig. 1.17. Psychrometric chart, showing how a condition may be defined by a number of variables.

pressure (Pa); the line CC^1 is a line of constant wet-bulb temperature very nearly parallel with lines of constant enthalpy (kJ/kg). There are thus at least seven variables which can each define the humidity of the air, and it is to be expected that each of the various humidity sensors will react more specifically to one of these variables than to the others. As with temperature sensors, it will be convenient to divide the various types of humidity sensor into electric and non-electric classes, but first it is necessary to deal with one method of humidity measurement which involves the measurement of temperature. Because these are of widespread importance they are usually given the title of psychrometers, indicating that humidity is measured by the combination of dry-bulb and wet-bulb temperatures.

1.2.1 Psychrometers

When a dry thermometer bulb is exposed to air in order to measure its temperature, it must reach a state of thermal equilibrium with the air by means of sensible heat transfer, by convective and conductive processes. If radiation from surrounding surfaces is eliminated then eventually the temperature sensor must take up the temperature of the surrounding air. If the surface of the sensor is wet, then the thermal equilibrium will now include a component of latent heat of evaporation of the water. Eventually the sensor will reach an equilibrium temperature θ_w, lower than the corresponding dry-bulb temperature θ. In this condition, the rate of sensible heat gain to the surface will balance the rate of latent heat loss by evaporation. The temperature θ_w is known as the wet-bulb temperature of the air. It can of course be measured by any of the types of temperature sensor which were described in Section 1.1. Equating these rates of heat loss and gain for unit surface area we can write:

$$(h_c + h_r)\cdot(\theta - \theta_w) = \alpha\cdot hfg\ (p'_w - p_s) \qquad (1.20)$$

where h_c = the heat transfer coefficient by convection at the water surface,

h_r = the heat transfer coefficient by radiation at the water surface,

α = the diffusion coefficient for water molecules between liquid water and free air, kg/s.Pa

hfg = the latent heat of evaporation of water at the temperature θ_w,

p'_w = the SVP of water at the temperature θ_w,

p_s = the actual vapour pressure of water in air at the given conditions.

Rearranging equation (1.20) gives an expression for wet-bulb temperature

$$\theta_w = \theta - (\alpha \cdot hfg/h_c + h_r) \cdot (P'_w - P_s). \qquad (1.21)$$

It has been empirically determined that the expression $(\alpha \cdot hfg/h_c + h_r)$ is equivalent to 1502/Pat where Pat is the barometric pressure. This gives the modified equation for wet-bulb temperature.

$$\theta_w = \theta - (1502/Pat) \cdot (P'_w - P_s) \qquad (1.22)$$

which applies for $\theta_w \geqslant 0^oC$ and for a sling wet-bulb, i.e. with an air velocity over the bulb of about 5 m/s.

The quantity $(1502/Pat) \cdot (P'_w - P_s)$ is the difference between dry-bulb and wet-bulb temperatures and is known as the wet-bulb depression. The wet-bulb depression is closely related to the relative humidity, increasing as relative humidity decreases. It is possible to construct a psychrometer using a pair of temperature sensors, one dry-bulb, one wet-bulb. If these are connected to a Wheatstone bridge in a differential arrangement, the output will be an approximate measure of the relative humidity. This can readily be established by plotting lines of constant wet-bulb depression on the psychrometric chart, when they will be found to follow reasonably well the lines of constant relative humidity.

In most cases, where the condition of room air is to be measured, it is convenient to use either a sling psychrometer or an aspirated psychrometer. The sling psychrometer consists of a wooden or plastic frame holding two identical mercury-in-glass thermometers, one of which has a muslin wick covering the bulb. The other end of the wick is immersed in water in a small plastic container. The frame has a pivoted handle so that the thermometers may be rotated in the air. The technique is first to ensure that the water is being carried along the length of the wick by capillary action and properly wetting the bulb. The psychrometer is then swung round for a minute or so and the two thermometer readings taken. The wet-bulb should be read first, as it starts to rise immediately. The process is repeated until a steady value of wet-bulb temperature is obtained. When swinging the psychrometer round, an airspeed of about 5 m/s should be aimed for, which means that it should rotate at about 4 rev/s. The aspirated or Assmann psychrometer also has a pair of mercury-in-glass or alcohol-in-glass thermometers, but the air is drawn over the bulbs at the correct speed by a fan driven either by a clockwork motor or an electric motor. The thermometers are fitted with double radiation shields. Because of these factors, the aspirated psychrometer is generally considered to be the most accurate psychrometric instrument in normal use. Psychrometers are not generally suitable for measuring rapidly varying conditions, only steady conditions, because their operation depends on an equilibrium being established between the wet bulb and its surroundings, which can take a few minutes. The sling psychrometer and the aspirated psychrometer are specified in British Standards 2842: 1975 and 5248: 1975 respectively.

1.2.2 Non-electric Humidity Sensors

The wet- and dry-bulb psychrometer was not the earliest type of hygrometer known. A hygrometer based on the change in weight of a ball of wool or hair seems to have been invented by Leonardo da Vinci about 1500, and it has been known for about 200 years that natural materials such as animal hairs change their length when the relative humidity of the surrounding atmosphere is changed. At present, three materials are in common use for humidity measurement or control: hair, gold-beaters' skin and nylon, also cotton is occasionally used. Gold-beaters' skin is the prepared outside membrane of the large intestine of the ox, used in gold-beating to separate the leaves of metal.

Hair has been used in hygrometers since the early nineteenth century. The length increases with increasing humidity and seems to be dependent on the relative humidity of the atmosphere surrounding the hair rather than on the other factors such as wet-bulb temperature or vapour pressure. The total change in length produced by a change in relative humidity from zero to saturation has been measured by various workers. Values reported vary from about 1.4% to 2.5% of the original length. This is obviously dependent on the source of the particular specimen used. Human hair seems to be favoured by manufacturers. The extension is also dependent on the surface properties of the specimen and on its cleanliness. The response is reduced by the presence of oil, grease, dust, etc., and hair humidity elements should be handled as little as possible. The calibration varies with time as atmospheric dust collects on the hair. The hair element should be cleaned and the hygrometer recalibrated at regular intervals. The change in length with relative humidity is not uniform, being more marked at low values of relative humidity, as shown in Fig. 1.18. In indicating instruments, non-uniform scale markings are used. In recording instruments such as thermo-hygrographs a non-linear mechanism is used so that uniform scale graduations can be employed. It has been found that if the hairs are rolled between steel rollers so that the cross-section is flattened, the total extension can be increased to about 4% of the original length. Unfortunately the mechanical strength is much reduced by this process of rolling.

Gold-beaters' skin is used by some manufacturers as a humidity sensor. Its advantage is that the total extension is about 6% of the original length, about 2 or 3 times as great as that of hair. The extension is also more uniform than that of hair. Unfortunately this material exhibits hysteresis effects and its properties are not constant between one specimen and another.

Cotton is used as the sensing material in humidity control by one manufacturer. It is unusual in that there is a decrease in length as the humidity increases, and the total amount of movement is small, only about 0.7% of the original length. The tensile strength is high compared with most other materials, and large forces can be exerted to operate a switch, etc.

It is known that a number of polymeric plastic materials are sensitive to variations in humidity, and nylon 6 is one which has been in use since about 1950 for indication and control. The total extension with relative humidity is about 4 or 5% of the original length, and the change is much more uniform than for hair, as can be seen from Fig. 1.18. As with hair, nylon seems to be sensitive specifically to relative humidity, rather than to any of the other psychrometric variables. The response at low temperatures (below about 10°C) is very sluggish, and hair is to be preferred for low-temperature work for this reason. Above about 10°C, however, nylon 6 is considered to have a superior performance to any of the other materials.

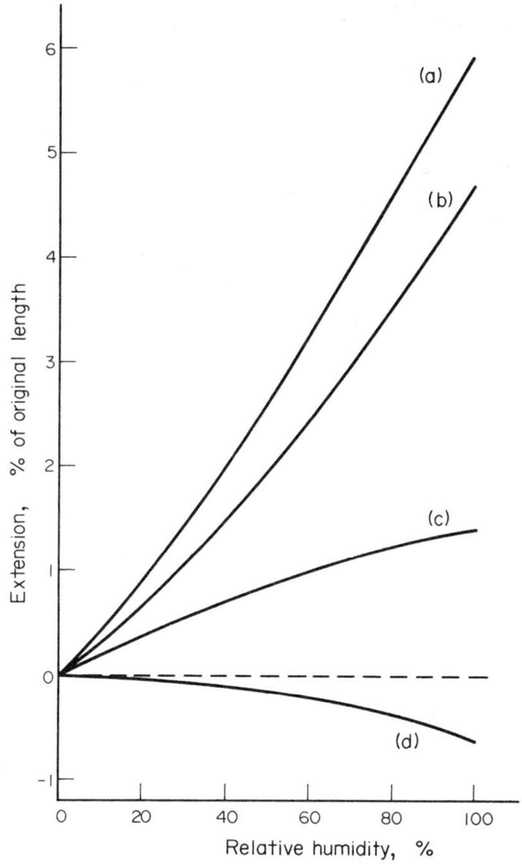

Fig. 1.18. Change in length of materials on increase in humidity:
(a) Gold-beaters' skin. (b) Nylon 6.
(c) Hair. (d) Cotton.

1.2.3 Electric Humidity Sensors

Although electrical thermometers can, of course, be used to sense wet- and dry-bulb temperatures in a psychrometer, a number of devices have been developed which give electrical output signals directly in response to the humidity of the air. Most of the earlier types depended on the hygroscopic properties of lithium chloride, and produced an output signal in the form of a change of electrical resistance over a number of decades. More recently, hygrometers have appeared which consist of electrical capacitors, whose capacitance value is dependent on the relative humidity of the surrounding air.

Because of the corrosive nature of the lithium chloride and calcium chloride salts used in the electrical sensors, it is necessary to use plastics for their construction, and, where metallic connections have to be made, platinum wire is often used. In one form, known as the Gregory element, the detector consists of a perspex former which holds four platinum-plated electrodes. A glass filament or yarn is wound round the electrodes and is impregnated with the hygroscopic salt. Another form, known as the Dunmore element, has a cylindrical perspex surface carrying a layer of LiCℓ and a bifilar winding of wires connected to plug-in terminals. In both these types a large change in electrical resistance is observed as the relative humidity of the surrounding air changes. The resistance could change by a *factor* of 10 for a change of about 30% in relative humidity. Unfortunately, this type of sensor is also very sensitive to temperature, so it is suitable only for applications where the dry-bulb temperature is known or is controlled. Alternatively, a compensating temperature sensor can be incorporated. Because of the large change in electrical resistance, a single sensor can cover only a narrow range of relative humidity, say 20%. In order to cover nearly the whole range, a set of sensors can be arranged, connected in parallel so that the sensors whose resistance is high do not affect the output value.

When these elements are used in Wheatstone bridge circuits, the bridge has to be fed with alternating, rather than direct, voltage to avoid polarisation effects. The bridge also has to be arranged to compensate for the effects of capacitance in the connecting cables.

One further type of humidity sensor which uses LiCℓ is known as the Dewcel or Dewprobe, shown in Fig. 1.19. As the name implies, it gives a measure, not of

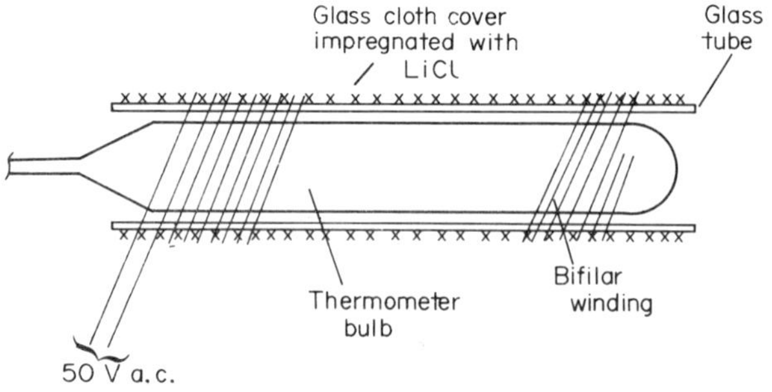

Fig. 1.19. Dewcel dew point element.

relative humidity, but of the dew-point of the air surrounding it. It consists of a glass tube covered with a wick of glass cloth, impregnated with LiCℓ. There is a bifilar winding of platinum wire over the surface, so that electrical current can travel between the wires only across the surface of the LiCℓ crystals. There is a temperature sensor inside the glass tube and the element takes up a temperature which is related, but not equal, to the dew-point temperature of the surrounding air. The bifilar windings are connected to an alternating voltage supply of about 50 V. This allows the surface of the element to maintain itself in equilibrium with the surrounding air. When the LiCℓ is dry its electrical resistance is high and almost no current is drawn from the supply; hence there is little or no heating effect and the hygroscopic salt is able to absorb moisture from the air. This causes the electrical resistance to fall sharply and a significant current is drawn from the supply. The heating effect of this current tends to drive off moisture from the LiCℓ and causes it to dry out so that its electrical resistance rises. Eventually the salt surface maintains itself at such a temperature that the Saturated Vapour Pressure of the LiCℓ solution is equal to the vapour pressure of the moisture in the surrounding air. Measuring the temperature of the element allows the dew-point temperature of the air to be found from a table or from a graph such as Fig. 1.20. This is a robust and reliable instrument and is not much affected by airborne dust deposits. The surface can be renewed by application of a saturated solution of LiCℓ in water.

Since about 1975 a humidity sensor working on the principle of electrical

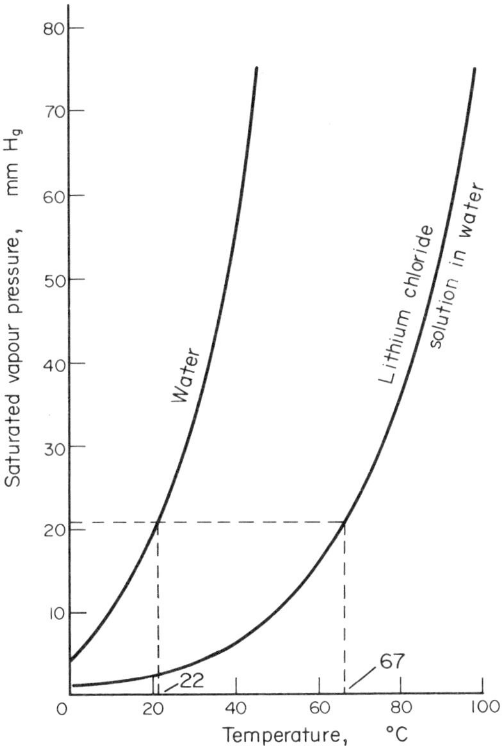

Fig. 1.20. Principle of the dewcel. As indicated a dewcel temperature of 67°C corresponds to a saturated vapour pressure of about 20.5 mm Hg and a dew point of about 22°C.

capacitance has become available. The capacitor consists of an amorphous polymer film, about 4 mm square, coated on both sides with metallic electrodes to provide electrical contacts. The electrodes must be thin enough to permit the passage of water vapour. Since the dielectric constant of water is high, absorption of water molecules by the polymer film causes a large change in the electrical capacitance. The change in capacitance as relative humidity increases from zero to 100% is about 30% of the capacitance when exposed to dry air. An electronic circuit is used to generate a voltage output directly proportional to the change in capacitance and hence to relative humidity. The effect shows very little hysteresis (less than 1% for 60% excursion of relative humidity) and very little temperature dependence (0.05% change in reading for 1°C change in temperature). The response time is short, being only 1 or 2 seconds at ordinary ambient temperature, but it increases rapidly as the temperature falls, being about 10 seconds at 0°C and about 40 seconds at -20°C.

CHAPTER 2

APPLICATIONS

2.1 FINAL CONTROL ELEMENTS

In practice, the variable which it is desired to control cannot be directly manipulated by the automatic controls. For room-temperature control the controller can often only alter the valve stem position and it is this which affects the room temperature. The valve stroke or the flow rate would be considered as the "manipulated" variable and it is usually required to make the changes in controlled variable a constant proportion of changes in the manipulated variable. Hence it becomes important to know what the flow characteristics and mixing characteristics are for various types of valve or damper. In order to convert the changes in controller output to physical movement of valve plug or damper blade an actuator mechanism is required. In general this will be electric or pneumatic, depending on the controller it is associated with, but it is possible to employ pneumatic actuators with an electronic detector and controller, and vice versa, although the latter would be unusual.

2.1.1 Pneumatic and Electric Actuators. Positioners

Actuators for valve or damper operation are usually either pneumatic or electric in principle of working. Electric actuators are driven by a small electric motor of about 20-30 VA power. The motor may be supplied either with 240 V or 24 V a.c. and runs at 1500 or 3000 rpm constant speed. It drives a reduction gear box to provide the low speed, high torque necessary for actuation. The final shaft torque is from about 2 to about 20 Nm. For fail-safe operation the shaft may be fitted with a heavy spiral spring. When power fails, an electromagnetic clutch, normally held in by the supply, is released and the spring is able to drive the valve or damper to a selected position. An alternative method is to provide a battery-operated power supply, also brought into operation by a line failure.

For actuation of modulating valves, the motors are of reversible type, but for operation of two-position valves such as small butterfly valves, unidirectional motors may be used, with associated limit switches to stop the motor when the valve has been driven either fully closed or fully open.

Two-position valves may also be operated by solenoids, giving rapid opening or closing of the valve. The solenoid operates on a sliding armature which is linked directly with the valve plug. These actuators are suitable only for the smaller sizes of valve and with limited pressure drop. They tend to be noisy and hence are not suitable for domestic applications.

Motor-operated valve actuators are relatively slow in operation, taking up to 2 minutes to drive over their full range. This is not a practical disadvantage for most applications. The effective speed can be further reduced by interruptors, as described in Section 3.2.

Pneumatic actuators, or pneumatic motors as they are also called, consist usually of a circular diaphragm on one side of which the control pressure is applied either directly from a controller or via a positioner (see Fig. 2.1). The other side of

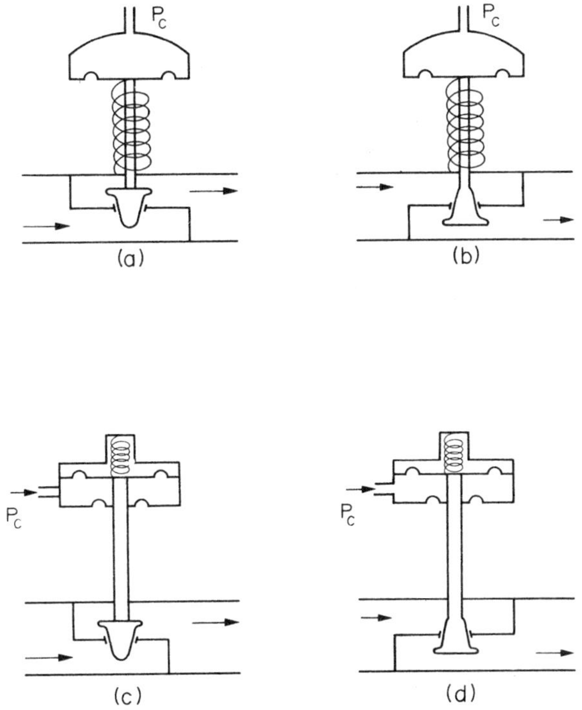

Fig. 2.1. Four possible arrangements of two-way valve and pneumatic motor (a) and (d) normally open, air to close, (b) and (c) normally closed, air to open.

the diaphragm bears on a helical spring, which provides an automatic return when air pressure reduces or fails. The movement of the pneumatic motor is rapid and very large actuating forces can be generated. For example, a moderately large diaphragm would be, say, 300 mm in diameter. The area is thus 0.071 m^2 and applying an air pressure of 1 bar gauge (10^5 Pa) to this diaphragm will generate a force of over 7 kN, enough to lift a mass of nearly three-quarters of a tonne against gravity. Such actuators can provide tight shut-off of single-seat valves against large pressures.

In cases where a long pipe connection exists between the controller and the pneumatic actuator, the movement of the actuator may be sluggish due to the long time taken for the pressure to rise or fall. In order to speed up the operation of the valve and to overcome effects due to friction in the glands and packing, a valve positioner may be used. This is a unit which senses the position of the valve stem and is able to draw on mains air pressure to ensure that the valve takes up the correct position as required by the control pressure. The principle of one design of positioner is shown in Fig. 2.2. The control pressure P_c is applied to the bottom chamber of the positioner and pushes against the diaphragm A. This causes the balance of forces on the T-shaped piece XYZ to be upset, and it rotates clockwise slightly. The valve at O is closed tightly and the valve at M opens slightly, admitting air into the upper chamber and increasing the pressure

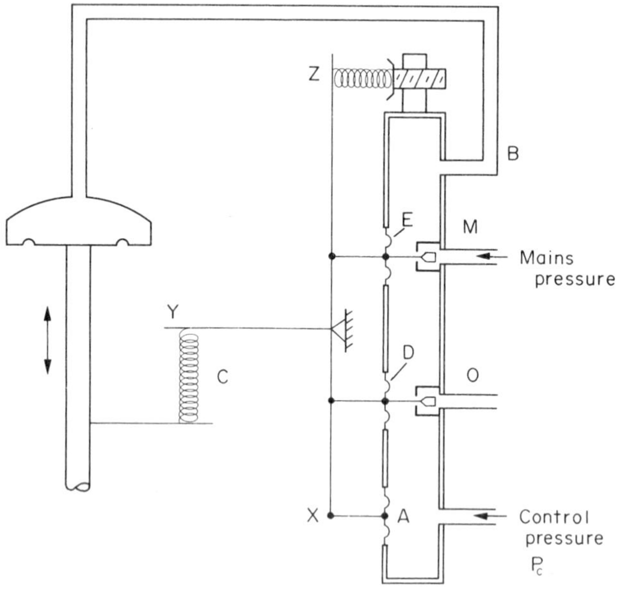

Fig. 2.2. Pneumatic valve positioner.

fed through the branch tube B to the valve diaphragm. If the valve stem does not move down the pressure will continue to increase until the valve is closed down. When the valve stem moves, the tension in the spring C increases and the balances of forces on the piece XYZ is restored. The two diaphragms D and E are equal in size and are placed an equal distance on each side of the pivot of the T-shaped piece XYZ: they allow movement, but the pressure fed to the valve motor has no effect on the balance of forces on XYZ. The minimum value of P_c at which the valve starts to move can be adjusted via the horizontal spring: two or more positioners can be used to make two or more valves operate in sequence or in unison from the same controller. A further advantage of the positioner is that the control pressure is feeding into a small-volume cavity rather than into the larger volume above the diaphragm. This produces faster response.

For damper actuation a much larger stroke is required than for valves. With electric motors a suitably long arm can be fitted to the final drive shaft to give a large movement. Pneumatic actuators for dampers usually take the form of a piston and cylinder arrangement, with a flexible rubber or neoprene sealing strip. The travel can then be up to 200 mm.

Self-operating actuators. Some self-contained actuators and controllers use the change in the controlled condition to provide the energy for the correcting action. A valve is actuated by pressure generated by a vapour pressure or liquid expansion system. The pressure drives the valve stem against a return spring. An example of these is the radiator thermostat used in car cooling systems, which utilises the expansion of a wax-filled capsule to operate a restrictor in the flow. Another type of self-operating controller is the constant air-flow regulator fitted to terminal units in air-conditioning systems. A spring-loaded flap is drawn across the duct by a high air pressure. A lower pressure will allow the flap to open, thus regulating the flow to a nearly constant value.

2.1.2 Valves

Control valves are essentially variable fluid resistances. They influence the flow of fluid by altering the free area presented to the flow. In building services applications the flow rate is usually an intermediate variable, modulated in order to control some temperature level by a varying heat-emission rate. The manner in which the flow rate varies with stem position is known as the valve characteristic, or inherent characteristic, since it is always measured in practice with a constant pressure drop across the valve. The characteristics usually available are: quick opening, linear, modified parabolic (also called characterized V-port), and equal-percentage. They are illustrated in Fig. 2.3.

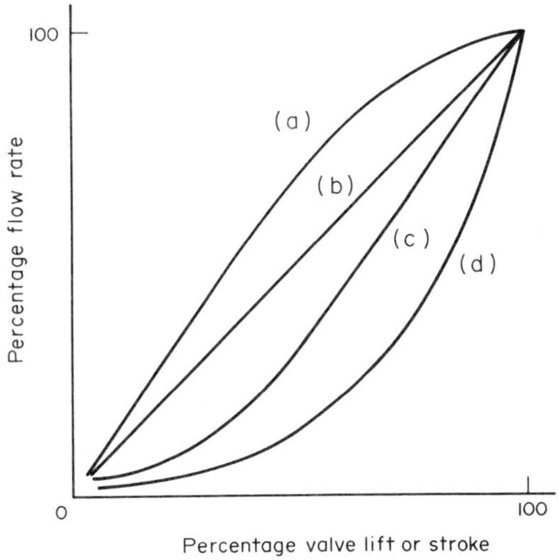

Fig. 2.3. Inherent characteristics of valves. (a) Quick opening, (b) linear, (c) parabolic, (d) equal percentage.

The quick-opening characteristic is not of much importance for modulating control purposes. For the other types, the flow rate follows approximately one of the functions given in equation (2.1).

Linear: $$Q = K_1 V, \qquad (2.1a)$$

Equal percentage: $$Q = K_2 \exp (K_3 \cdot V), \qquad (2.1b)$$

Parabolic: $$Q = K_4 V^2 \qquad (2.1c)$$

where Q = flow rate, m^3/s or percent of maximum,

V = valve stroke, mm or percent of maximum

and K_1 to K_4 are constants for the given valve.

Note that equation (2.1b) differs slightly from that given in BS 5384: 1977 (Section 3.3.1), where there is a printing error.

Figure 2.4 (a) to (f) illustrates the construction of the different types of valve.

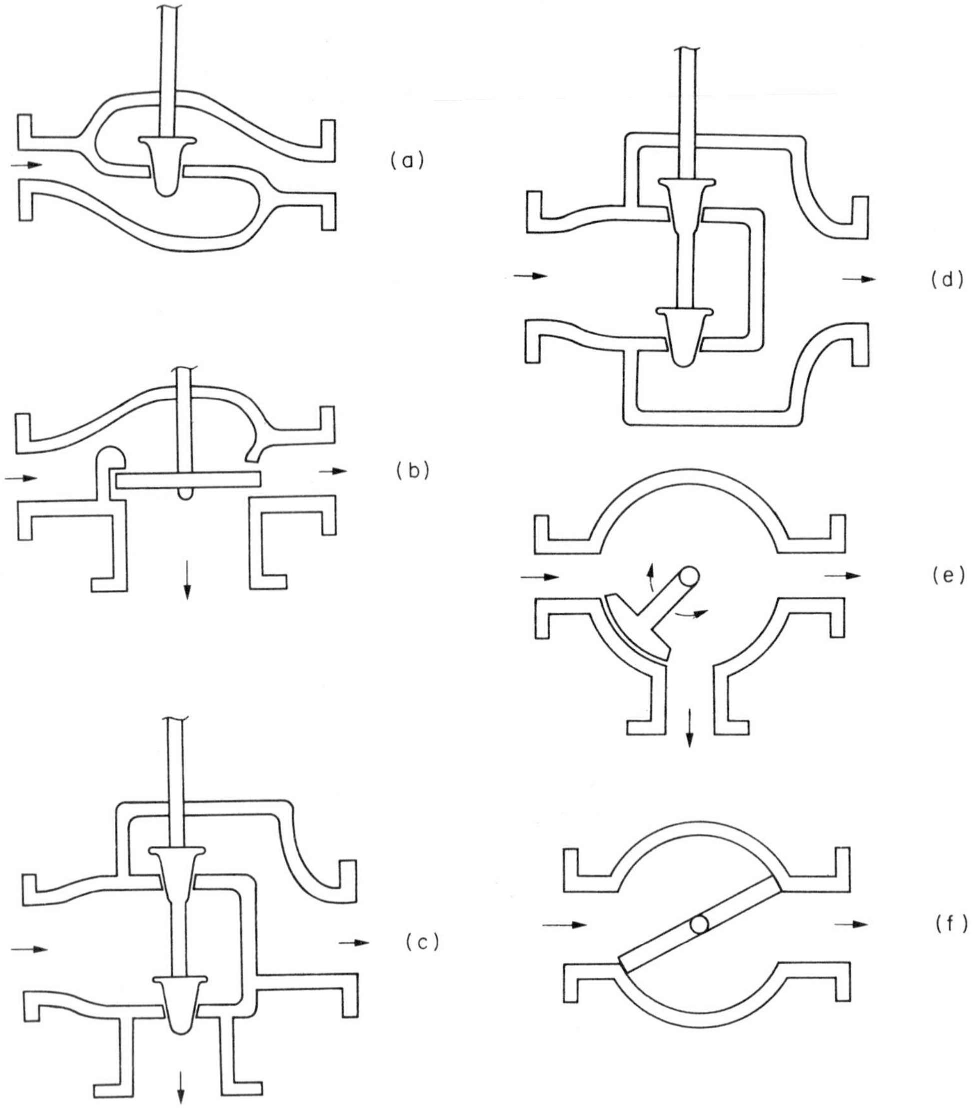

Fig. 2.4. Valve types: (a) single seat two way; (b) single seat three way; (c) double seat three way; (d) double seat two way; (e) shoe or slipper valve, three way; (f) butterfly valve.

With single-seated valves the whole of the pressure differential across the valve is exerted across the plug. The actuator must be able to generate sufficient force to close off the valve against the pressure. It is important that, in a two-port valve, the direction of flow is such as to oppose the closure of the plug. Otherwise, when the valve is nearly closed the differential pressure will cause the plug to snap shut and perhaps to oscillate with consequent noise, vibration and rapid wear. In double-seated valves the forces are nearly evenly balanced between the two plugs and much smaller actuating forces are needed. On the other hand, a tight shut off cannot be obtained with double-seated valves and this is a limitation on their use in some cases, such as steam. The design of the butterfly valve permits a much lower fluid resistance when fully open than is possible for other types. The inherent characteristic is not suitable for most control applications. Shoe valves or slipper valves are simpler to manufacture than most other types and the internal ports can be shaped to give a required characteristic. The rotary motion of the shaft is convenient for electric actuators. The leakage rate is often higher than for plug or skirt-type valves.

The flow-handling capacity of a control valve is normally denoted by the flow rate due to a unit pressure difference across the valve *when the valve is fully open.* This flow rate is termed the flow coefficient or velocity coefficient. When the flow rate is measured in (US) gallons per minute and pressure in lbf/in^2 the coefficient is designated C_v. When the flow rate is measured in m^3/h and pressure in kgf/cm^2, as in continental practice, the coefficient is designated by K_v. For a given valve, the value of C_v is about 17% larger than that of K_v, and runs approximately from a minimum of 0.2 for 12 mm valves up to about 200 for 120 mm valves. The actual value of C_v depends on the valve type and trim as well as on the nominal size.

An alternative definition of a flow coefficient has been given in British Standard 4740: part 1: 1971. This defines a flow coefficient A_v, based on the flow rate in m^3/s for a pressure drop of 1 Pa across the valve. The conversions between A_v, C_v and K_v are as follows:

$$A_v = 28.9 \times 10^{-6}\ C_v \qquad \text{(Imperial gallons)}, \qquad (2.2a)$$

$$A_v = 24.0 \times 10^{-6}\ C_v \qquad \text{(US gallons)}, \qquad (2.2b)$$

$$A_v = 28.0 \times 10^{-6}\ K_v \qquad (m^3/h). \qquad (2.2c)$$

The factor of 10^{-6} is perhaps confusing and can be eliminated if the metric units used are litre/sec and bar. The multiplying factors then become 0.289, 0.24 and 0.28 respectively.

Valve authority. Consider the case of a two-port valve which is required to influence the flow through a system by throttling, i.e. by increasing its fluid resistance to cut down the flow. If the pressure drop across the valve in a simple circuit such as that shown in Fig. 2.5(a) is only say 1% of the total then a doubling of the fluid resistance when the valve is half closed will have a negligible effect on the flow rate. On the other hand, if the pressure drop across the fully open valve is 90% of the total then the flow will be strongly influenced by the valve position. This introduces the concept of valve authority, N, which is defined as follows.

$$N = \frac{\Delta P_1}{\Delta P_1 + \Delta P_2} \qquad (2.3)$$

where ΔP_1 = pressure drop across the valve when fully open,

ΔP_2 = pressure drop across the remainder of the circuit.

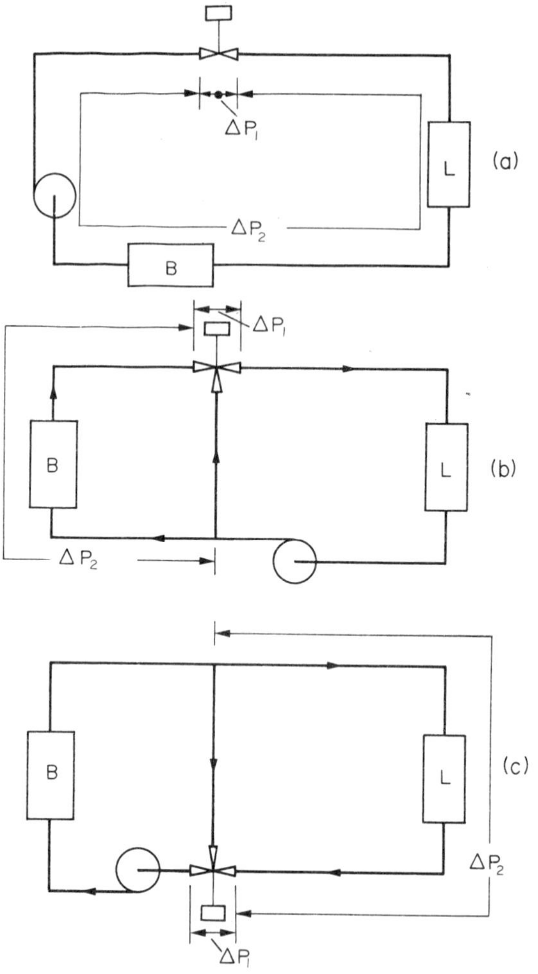

Fig. 2.5. Definition of valve authority: (a) simple throttling application; (b) three-port mixing application; (c) three-port diverting application. In each case, valve authority $N = \Delta P_1/(\Delta P_1 + \Delta P_2)$, B = Boiler, L = Load.

Note that for three-port valves "the remainder of the circuit" means only that part of the circuit in which the flow rate is being varied by the control valve.

Installed characteristic. It is important to distinguish between the inherent characteristic and the installed characteristic, as affected by the authority. If one considers a valve with an ideal linear inherent characteristic, such a valve installed in a system with a small authority will have to close nearly completely before it has much influence on the flow rate. Figure 2.6 indicates the effect of authority on the installed characteristic for such a valve.

Because too small a valve authority will give too little control over the flow and a non-uniform characteristic, and too large a valve authority will increase the pumping power requirement, some compromise is necessary. A value of 0.4 (or 40%) at least is usually satisfactory, but a larger value is desirable.

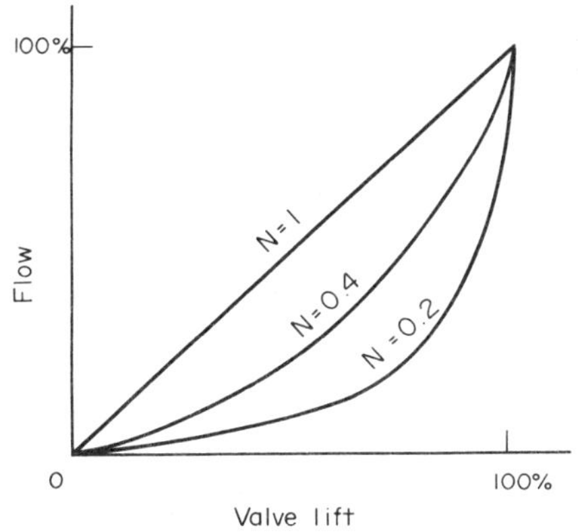

Fig. 2.6. Effect of valve authority N on installed characteristic-linear valve.

Valve rangeability and turn-down ratio. With most control valves the fully closed position does not give a tight shut off. There is still some flow through the valve, termed let-by or leakage. At some point before the fully closed position there is no longer proper control over the flow, and the nominal characteristic is no longer adhered to. The rangeability of a valve is defined as the ratio of the maximum controllable flow to the minimum controllable flow. When a valve is installed in a pipeline the maximum flow depends on its authority. The turn-down ratio is defined as the ratio between the maximum normal flow and the minimum controllable flow. If the pressure drop across the valve is small then the turn-down ratio will be much less than the rangeability. Values of rangeability are from about 100 for a single-seated valve in new condition down to about 15-20 for a double-seated valve after some time in service.

Choice of valve characteristic. The choice of characteristic usually lies between a linear and an equal-percentage type. Some general indications for guidance are as follows:

A linear characteristic is chosen when a small rangeability is acceptable, the valve authority is high and the system pressure is constant.

An equal percentage characteristic is chosen when the system pressure may vary widely, when the valve authority is relatively low and when there is uncertainty about the flow conditions.

Steam valves. It has previously been assumed in this chapter that the medium whose flow is to be controlled is an incompressible fluid such as water. Where a compressible fluid such as steam is to be controlled, some special considerations are introduced.

The flow of steam through a restrictor of area A can be written as

$$\dot{m} = K \cdot A\,(P_1 - P_2) \qquad (2.4)$$

where $\dot{m}$ = mass flow rate, kg/s,

A = orifice area, m^2,

P_1, P_2 = upstream and downstream pressures (absolute), Pa,

K = a constant.

With a steam valve the area A is reduced as the valve closes down. The downstream pressure P_2 reduces as A reduces so that the pressure difference P_1-P_2 increases. These two effects largely cancel and the flow remains constant, until P_2 drops below approximately 60% of P_1. At this point the flow may be controlled.

The steam valve should be sized on the basis of the flow of steam required and the upstream pressure P_1. The pressure drop across the valve may then be taken as 40% of P_1, and the valve chosen from manufacturers' tables.

2.1.3 Dampers

Dampers are employed in air-flow systems to provide a variable fluid resistance and control flow rate in the same way as control valves in liquid flow systems. In small duct systems single-leaf butterfly-type dampers are often used for balancing purposes, but their only virtue is cheapness as their flow characteristics are very non-linear. A small movement from the fully closed position gives a very large change in flow rate. For automatic control purposes rectangular dampers with a number of parallel blades or leaves are used, each blade rotating on its axis like a venetian-blind arrangement. There are two alternative methods of arranging the blade rotation, either parallel where all the blades have the same angle to the airstream and rotate in the same direction, or opposed, where alternate blades rotate in opposite directions. Opposed blade dampers do not tend to act like turning vanes and so do not direct the air flow to one side of the duct. Parallel blade dampers do have this effect and can cause stratification. The linkage arrangements are more complicated for the opposed blade type.

Dampers possess inherent characteristics, installed characteristics and authority in the same way as do valves. Parallel blade dampers generally have a more uniform inherent characteristic than do the opposed blade type, but the installed characteristic of opposed blade dampers is better than that of the parallel type, taking the authority to be about 10%.

2.2 DOMESTIC HEATING CONTROLS

It is considered that the domestic system to be controlled consists of a central boiler serving either LPHW radiators or convectors with pumped circulation, or a warm-air system with a fan. In domestic premises, the expenditure on controls is usually very limited but the requirement for economy is very direct. The simplest possible control is to have a time switch or programmer which switches on the space heating and domestic hot-water heating at predetermined times. The space heating is switched on by energising the circulating pump motor and the domestic hot water by opening a butterfly valve in the gravity circulation primary circuit. The programmer usually carries selector switches so that heating and hot water can be separately chosen to be on constantly, under time-switch control, or off. When both services are off the boiler should be switched off automatically, and on only when one or both of the services are on. There is no room thermostat in this system and space temperature control basically depends on the boiler thermostat and its temperature setting, and on the sizing and balance of the heat emitters. The valve in the primary domestic hot-water circuit can be closed off by a thermostat installed on the cylinder (about three-quarters of the way up the cylinder is a usual position). This arrangement is most suitable for gas or oil-fired boilers, but it can be used with solid-fuel boilers, where some permanent load such as a towel rail exists to absorb the minimum firing rate. The user will adjust the boiler thermostat high in winter and low in summer, providing a sort of manual outside compensator. This will affect the temperature of hot water at the taps. There is also no provision for individual room-temperature control or zoning, and there is the possibility of the boiler being affected by low return water temperatures.

A more sophisticated system than the above rather crude arrangement will incorporate a room thermostat to control space temperature and a three-way mixing valve to permit the domestic hot-water temperature to be largely independent of boiler temperature. The siting of the thermostat can be a difficult matter. There is often no single position where the temperature is representative of the whole house. With pumped heating circulation and gravity circulation in the hot-water cylinder, the space thermostat can conveniently be arranged to control space temperature by switching the pump on and off.

In systems where there is a pumped primary to the domestic hot-water cylinder both functions can be made available separately by a three position motorised valve, or by individual small motorised valves.

Individual room-temperature control can be achieved most simply by fitting thermostatic radiator valves to heat emitters. This avoids fuel waste due to overheating when, for example, solar gains are sufficient to offset heat losses in some rooms. In this case a bypass between boiler flow and return may be required.

In addition to the time switch or programmer which is often provided with the boiler, larger domestic systems may have simple outside compensator control with provision for morning boost and night setback (see Section 3.3.2).

The control systems described above, except for the thermostatic radiator valves and outside compensators, can also be applied to warm-air systems. Individual room control in such systems may be obtained manually with adjustable dampers in each room or automatically with thermostats controlling motorised dampers.

2.3 BOILER AND CHILLER CONTROLS

In small to medium-size boiler installations (say less than 100 kW) there will usually be only a single boiler with its individual controls. Larger installations generally have multiple boilers which are arranged to be fired sequentially as the heating load increases. Special control arrangements are necessary for multiple boilers to ensure safety and accuracy of control. Burners may be of three types: on/off, high/low/off and modulating.

Figure 2.7 shows the control arrangements for a single boiler system. This applies where only one boiler at a time is sufficient to meet the load: there may

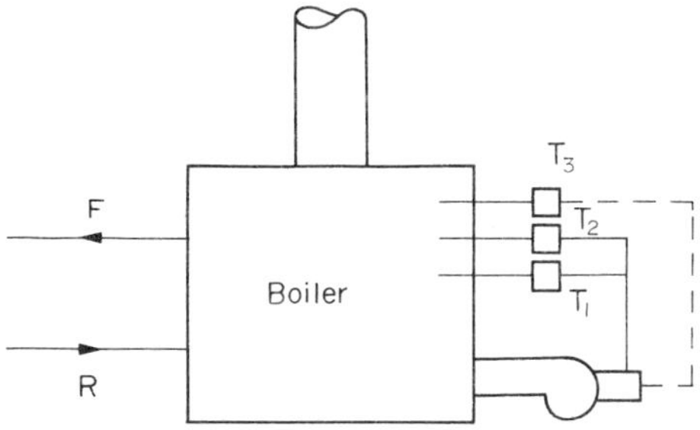

Fig. 2.7. Control of a single oil or gas-fired hot water boiler with on/off or high/low/off or modulating burner. T_1 = boiler temperature detector; T_2 = high-temperature safety cut-out; T_3 = high/low flame thermostat (required only for high/low/off control).

also be another standby unit. The boilers are normally delivered with their own packaged control systems. There must always be an independent high-temperature limit cut-out, which will stay cut-out once operated until reset by hand. The same cut-out can operate an alarm bell or lamp when it is activated.

Where multiple boilers are used they are fired in a predetermined sequence as the load increases, using a step controller fed by a proportional controller sensing return water temperature. On decreasing load they may be arranged to shut down in the same sequence as for start up, or in reverse sequence. There may be manual or automatic change of the "lead" boiler in the sequence to ensure even usage of all boilers. The individual control arrangement in each boiler is generally the same as in a single boiler installation, with the addition of the step controller for sequencing, as shown in Fig. 2.8. Each boiler may be on/off, high/low/off or

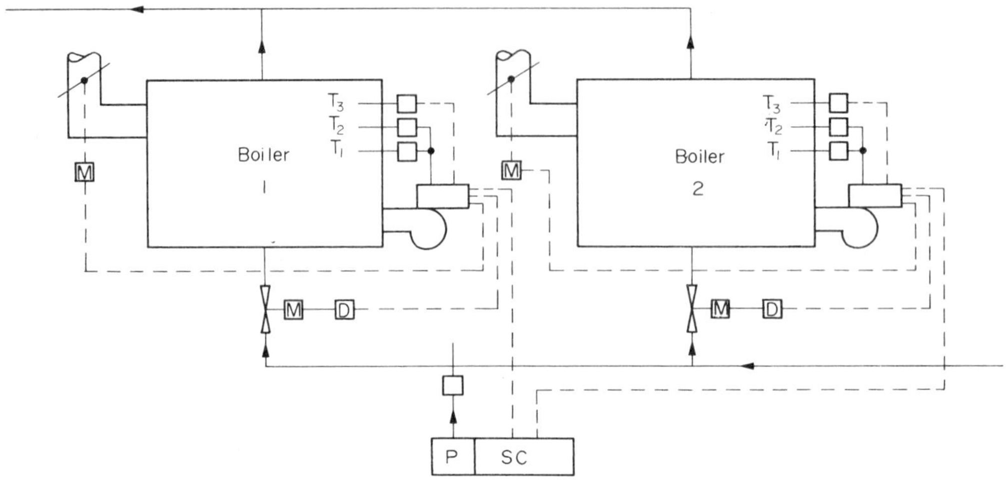

Fig. 2.8. Control of multiple boilers via step controller. T_1, T_2 and T_3 are as Fig. 2.7. D: delay timer to avoid overheating on shut down. P,SC: proportional controller and step controller. M: motor actuator.

modulating. When an individual boiler is shut off a motorised valve in the return is closed after a delay of 2 or 3 minutes to avoid cut out on the high-limit thermostat. It is desirable also to arrange for a motorised flue damper to close, conserving the heat stored within the metal of the boiler. A corresponding delay must be arranged before the burner is fired to allow time for the flue damper and the return water valve to be driven open. A reverse return arrangement for the piping is desirable to ensure that the water-flow rate is the same through each boiler.

Multiple boilers and chillers may also be piped in series so that the same water passes through each in turn. This can provide a simpler piping layout but the pumping power requirement is greater. The dangers of over-heating in boilers and subcooling in chillers are removed by using the series arrangement. With both the single and multiple boiler arrangements, an outside compensator system may be used to adjust the boiler-flow water temperature in accordance with outside temperature.

2.4 AIR-CONDITIONING CONTROLS

An example of the control system for an air-conditioning system is described in Section 3.3.2, but much simpler systems are often used. These may provide temperature control only, humidity control only or so-called "full air-conditioning", with control of both temperature and humidity. Modulation of dampers to adjust the proportion of air which is recirculated can also be used, allowing some saving on energy consumption of the plant.

Figure 2.9 shows the control arrangement for a simple heated air system, using sequential control of air recirculation and of heating. Air filtration, although

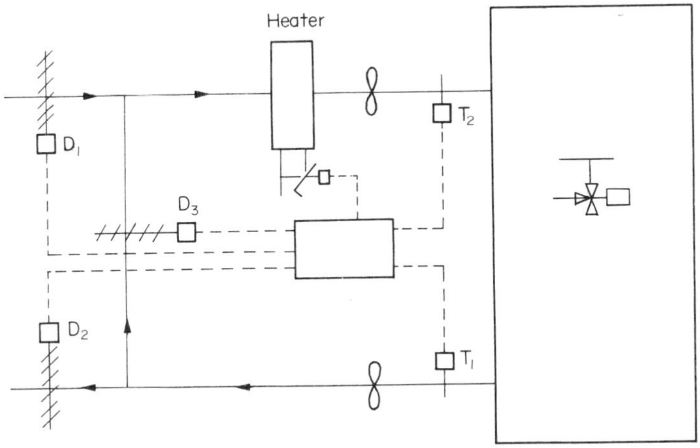

Fig. 2.9. Simple heating system with modulating dampers and heater battery. D_1, D_2, D_3: inlet, exhaust and recirculation dampers, T_1: exhaust air sensor, T_2: supply-air-limit sensor.

not shown, is an essential component of the system. In cold-weather conditions the dampers are modulated to their extreme positions, with minimum fresh air and maximum recirculation. The valve on the heater battery provides control of the supply air temperature to offset heat losses. Note that the overall control is provided from the space temperature detector. This may conveniently be sited in the extract air duct. A sensor in the supply air acts as a low limit detector. When the outside temperature rises and heating is no longer required, the recirculation damper starts to close and at the same time the inlet and extract duct dampers are opened. This continues until the extreme condition, of full fresh air operation with no recirculation, is reached.

In addition to the simple system in which air treatment is applied at a central plant and the conditioned air is distributed to the spaces served, a number of alternative terminal arrangements are possible. Some of the more common systems are described below.

Dual duct systems. The basic principle of these systems is to distribute two airstreams, one cold and one hot, in separate ducts to terminal units. In the conditioned space are mixing boxes to which the two ducts are connected. Each box or group of boxes mixes the two airstreams in proportions suitable to the room cooling or heating load, as controlled by a room thermostat. If the air conditions in the hot and cold ducts are represented by two points on the psychrometric chart, the mixed condition will fall on a straight line jo[illegible] the two points. An actuator in each mixing box operates dampers [illegible] and cold supplies to give the required mixed supply air condition. An automatic constant volume regulator ensures that the total flow rate from the box remains constant, and hence the air-distribution pattern in the room. Room thermostats of the type described in Section 3.4 may be used to control the modulating motors in the mixing boxes. It is important that all the boxes in one room or conditioned space are controlled by one thermostat. Individual control, where one thermostat can be affected by a terminal unit controlled by another thermostat, will produce unsatisfactory conditions due to simultaneous heating and cooling.

The hot and cold deck temperatures are reset depending on outside temperature, according to a predetermined schedule. This type of system can give accurate control, but is rather wasteful of energy since there is unavoidable mixing of hot and cold air streams. Separate and individual control of both temperature and humidity are not really possible.

Induction unit systems. In these systems the latent heat, dehumidifying function necessary in full summer air conditioning is separated from the sensible heat, cooling function. The central plant supplies to each room primary dehumidified air and water at a suitable temperature. The primary air is delivered through a set of nozzles at rather high velocity so that about 4 times its own volume of room air is entrained. The mixed air passes over a finned tube heat exchanger through which the water supply passes, giving local temperature control. In winter the heat exchanger provides heating, in summer it gives secondary cooling. The changeover from cooling to heating requirement at the induction units can give rise to some inconvenience, especially in the European climate. As well as changing from hot-water to cold-water circulation in a two-pipe system, it is necessary to change over from direct action to reverse action at the thermostat. This has given rise to three-pipe and four-pipe induction units, each containing a heating coil and a separate cooling coil. Each unit is fitted either with two control valves, one for each coil, or with a damper arrangement to give the same effect. The necessity of special arrangements and the delay when changeover occurs between heating and cooling is thus eliminated. The four-pipe system is to be preferred on grounds of energy economy. As with dual-duct systems, it is important to ensure that units serving the same space are all controlled by the same sensor or thermostat.

Variable air-volume systems. In order to meet changes in cooling or heating loads on a room, an alternative to supplying air at constant flow rate and variable temperature is to supply at constant temperature and variable flow rate. Reduction in load will result in reduced running costs by saving on fan power, provided that the variation in flow rate is *not* achieved by using simple dampers to throttle the flow. Special solid-state units are available for modulating the motor speed, and inlet vanes on centrifugal fans or variable-pitch blades on axial fans can be employed. As with the systems described above, a central dew-point-controlled plant may be used, and individual temperature detectors are arranged to reduce the volume of air from terminal units. Variable air-volume systems are most commonly applied to those buildings where there is a continuous requirement for cooling.

2.4.1 Zoning

Although central plant for heating or air-conditioning has to be sized with regard to the total load occurring over the whole building, it is not true that all parts of the building experience their peak heating or cooling loads simultaneously. This is especially the case for cooling loads, which are strongly determined by solar intensity on particular faces of the building. Peak cooling loads will occur early in the day for east-facing rooms and late in the day for west-facing rooms. South-facing rooms will experience maximum cooling loads around noon (in the Northern hemisphere). Heating and cooling loads are also influenced by patterns of occupancy, internal heat gains and by the required environmental conditions in individual spaces.

Where sections of the building can be identified as having individual requirements differing in time-pattern from adjacent sections, then the plant and controls must be arranged to permit separate control of each section. The simplest arrangement is, of course, to split the building according to the aspect, i.e. east/west. In tall buildings a split into upper and lower zones may be necessary because of the greater exposure of upper storeys. Central areas of buildings, remote from the perimeter, are not much influenced by external weather conditions and zoning of such areas is not usually necessary. Corner rooms which may appear to belong to two zones at once need special consideration to avoid unsatisfactory conditions, possibly due to simultaneous heating and cooling.

2.5 SOLAR COLLECTOR SYSTEM CONTROLS

In order to consume the minimum amount of energy in operation a solar collector panel should be located at a lower level than the storage tank which it feeds, so that gravity circulation can be used to provide flow through the panel. During periods when little or no solar radiation is available there would then be no tendency for heat to be extracted from the store and lost: the circulation would stop automatically as it does in a domestic boiler/cylinder arrangement with a gravity circulated primary. Unfortunately it is rarely practical to arrange the solar collector system in this way. Usually the collector panels are most conveniently mounted on the roof of the building and gravity circulation is a disadvantage. Pumping is necessary and a control system must be used to ensure that pumping occurs only when the temperature at the collector panel output is higher than that at some representative point in the storage tank. The control circuit is usually of a simple Wheatstone bridge form, with two sensors, one for the panel output temperature and one for the storage-tank temperature. Platinum resistance elements or thermistors may be used. The bridge is adjusted so that when the panel temperature exceeds the storage-tank temperature by 0.5 to $1^{o}C$ the pump is switched on via a contactor. Otherwise the pump is off and a low-resistance butterfly valve is closed to prevent gravity circulation. A time delay may be incorporated to ensure that the pump is not started too frequently, and safety limit detectors to guard against overheating in summer and frosting in winter.

2.6 BUILDING AUTOMATION SYSTEMS

Since the dramatic fall in cost of digital computing power which became evident from about 1970, it has become economically and technically feasible to employ a digital computer to carry out the functions of a large number of individual conventional controllers. In direct digital control, which was first applied to chemical engineering process plants, the outputs of a large number of sensors are scanned sequentially. Each output is compared with a stored value of set point for that sensor and any corrective action necessary is computed and applied as an output to the appropriate final control element. If the number of points to be scanned and the scanning rate are such that each point can be checked at intervals short compared with the associated plant time constants, then the effect of continuous control at each point is produced. The adjustment of set points and controller parameters can conveniently be done centrally under the supervision of a special computer program. Variables which are not under control can be scanned to provide a check on proper operation of the plant, safety conditions, fire, etc. The dependence on the reliable working of a single computer system can be eliminated by providing a second computer duplicating the operations of the first. Some systems use three computers working in parallel and control or alarm actions are taken on a "majority vote" principle.

This type of control has become quite well established in the process industries where the costs involved could easily be justified economically by quite small improvements in reliability and product quality. It is only relatively recently that digital controllers and supervisory systems have spread to the building services field. Instead of applying digital systems to the control functions of services, it is usual for building automation systems to be largely employed in supervision, security and alarm functions. Control has been mainly confined to switching functions, replacing time switches and optimum start controllers and having much greater flexibility. More recently such systems have also been used for continuous feedback control. Dependence on a single computer processor is reduced by the use of remote stations each with its own microprocessor and hence capable of some independent operation, but under the control of a central computer for higher-level functions. The installation of a building automation system can be justified on the basis of the wide range of functions and services which are available, related not only to heating and air conditioning but also to monitoring and control of lighting, electric power, communications, lifts, fire detection and alarm, fire fighting, cold-water services, plant management for energy conservation, maximum demand monitoring of electricity use and planned and programmed maintenance schedules.

2.7 PLANT AND ROOM TRANSFER FUNCTIONS

In order to apply much of the control theory presented in Chapters 5 and 6, it is necessary to specify the dynamic behaviour of the controlled system in transfer function terms. Unfortunately, there is little well established data available for many of the components in building services systems. The thermal dynamics of building structures themselves are probably the least well understood of the components in the system.

For room-temperature dynamics, a transfer function of the form

$$\frac{K \cdot \exp(-Ls)}{1 + Ts}$$

has been recommended. The value of K is related to the steady-state heat-loss rate of the room and the values of the pure delay L and the time constant T are determined from the values of room volume and air-flow rate.

For pipes and ducts, a simple approach would be to assume a heat loss proportional to the temperature difference between the fluid and the surroundings and a pure delay or dead time dependent on the flow velocity and distance, giving the form

$$\theta_1 \cdot \exp(-L_1 s) \cdot \exp(-L_2 s)$$

where θ_1 is the temperature of the fluid entering the pipe section, L_1 is the pure distance-velocity lag and L_2 is a temperature-decay constant dependent on the specific thermal capacity of the fluid and the heat-loss properties of the duct or pipe wall.

The dynamic responses of a number of commercially available room thermostats have been measured. They were found to be modelled with fair accuracy by a short delay time and a single time constant, giving the transfer function

$$\frac{\exp(-Ls)}{1 + Ts}$$

where L was measured to be 30 to 40 seconds and T was between 700 and 1200 seconds for step changes in air temperature. The values for step changes in radiation were about 10% higher. The steady-state temperature responses to air temperature and to radiant temperature could be expressed by a simple equation:

$$\theta_t = Y \cdot \theta_a + (1-Y)\theta_s \quad (0 < Y < 1) \tag{2.5}$$

where θ_t = thermostat sensor temperature,

θ_a = air temperature,

θ_s = mean surface temperature of surroundings,

Y = a factor expressing the relative sensitivities to air and radiant temperatures.

Generally thermostats are found to fall into two classes, those with a pierced casing, giving a "ventilated" effect, and those with a casing which is not pierced, giving an "enclosed" effect. Obviously, the ventilated type is more responsive to air temperature than is the enclosed type. Mean values for the parameter Y are 0.9 for the ventilated type and 0.67 for the enclosed type.

For temperature sensors generally, the time constants can be predicted by the method described in Section 5.2.

For room air heaters of various types and for hot-water radiators with various flow-rates and piping arrangements, Tables 2.1 and 2.2 give values of time constants. These are adapted from Reference 1 (Adams and Holmes, 1977).

TABLE 2.1 Time constants of room air heaters
(from Adams and Holmes, 1977)

	Time constant (minutes)	
	Heating up	Cooling down
Portable convector	2 ± 0.5	6 ± 1
Sill-line convector	0.5 ± 0.25	2 ± 0.5
Tube heater	9 ± 1.0	9 ± 1
Heated floor	180 ± 30	240 ± 60
Oil-filled radiator	17 ± 1.0	25 ± 1.0
Hot-water radiator	See Table 2.2	30

TABLE 2.2 Time constants of hot-water radiators for various water-flow rates (from Adams and Holmes, 1977)

	Time constant (minutes)	
Water-flow rate (kg/s)	Single-pipe system	Two-pipe system
0.014	5.5	8.0
0.021	4.0	5.3
0.032	2.9	3.6
0.048	2.3	2.7
0.068	1.5	1.9

CHAPTER 3

CONTROLLER MECHANISMS AND CIRCUITS

3.1 PNEUMATIC CONTROL - AIR SUPPLY

Pneumatic control systems use compressed air to supply energy for the operation of sensors, motors, actuators, etc. Compressed air is not normally available in most buildings as a matter of course, in the way that single-phase electric mains power is available. The air supply has to be specially installed where pneumatic controllers are to be used. Plant room space has to be provided for the compressor and the purchase and installation of the compressor and distribution pipework will add to the total cost of the system. Since a number of the components used in a pneumatic system are simpler and cheaper than the corresponding components for electrical systems, the cost of purchasing and running the compressor may be justified if the installation is above a certain size.

Figure 3.1 shows the main components in the compressed-air supply system. The compressor draws atmospheric air in through a filter and supplies it to the tank or

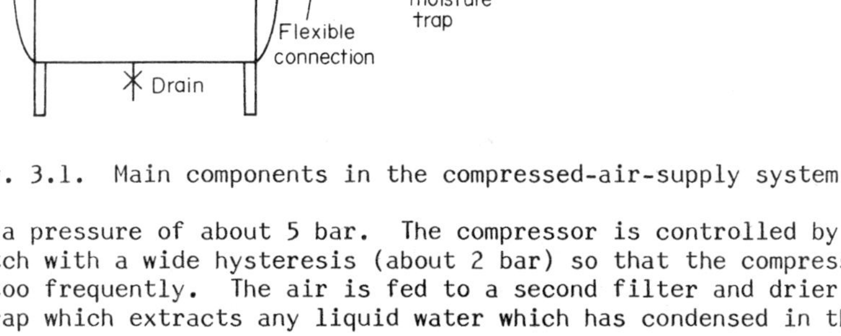

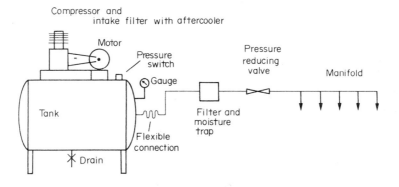

Fig. 3.1. Main components in the compressed-air-supply system.

reservoir at a pressure of about 5 bar. The compressor is controlled by an on/off pressure switch with a wide hysteresis (about 2 bar) so that the compressor is not switched on too frequently. The air is fed to a second filter and drier unit with a moisture trap which extracts any liquid water which has condensed in the air and would be carried along with it. The pressure is reduced to the working level of about 1.2 bar by a pressure-reducing valve and fed to a pipe manifold from where it can be distributed by narrow-bore copper or plastic pipework to the various controllers and sensors. All but the smallest systems should have dual compressors with automatic switch-on of the standby unit on low pressure in the duty unit.

3.1.1 Continuous-bleed Proportional Controller

Since the basic variable in pneumatic systems is the air pressure, it is necessary to have some means whereby the sensors can directly affect the air pressure at a point. One such method is called the flapper and nozzle system, shown in Fig. 3.2. This system is called continuous-bleed because air is continuously escaping from the nozzle. The distance of the flapper away from the nozzle determines the fluid resistance of the flapper/nozzle combination, so this forms a variable restriction to the air flow. There is also a fixed restriction which the air must pass through. The fixed restriction is only about 0.2 mm diameter, and the nozzle diameter is about twice this. Obviously, as the temperature rises and the flapper tends to close up the nozzle, the pressure drop across the fixed restrictor will fall and P_c will rise because the main air pressure is constant. As the temperature falls the flapper will move away from the nozzle, the air flow will increase and the pressure drop across the fixed restrictor will increase. The control pressure P_c will fall. The relationship between P_c the control pressure and the flapper-nozzle clearance

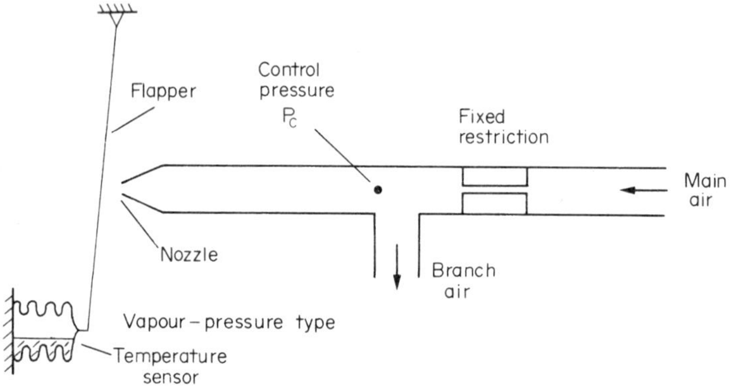

Fig. 3.2. Basic flapper and nozzle mechanism.

is shown in Fig. 3.3. It can be seen that the curve is rather S-shaped, starting at zero slope for zero clearance, increasing in slope to a maximum and then the slope decreasing to nearly zero again. There is a central section where the slope is reasonably constant and controllers and actuators are normally designed to operate within this pressure range of 0.2 to 1.0 bar. It is worth noticing that this pressure change is produced by a change in clearance of only about 15 microns (1 micron = 10^{-6} m). Such a small movement could easily be produced by thermal expansion of the mechanism on change in temperature, or by wear at the pivots, etc., so a control mechanism as shown in Fig. 3.2 would be almost useless. In order to make this very high gain element useful and reliable, it is necessary to stabilise it by the addition of negative feedback. This is a principle which applies to almost every continuous controller, pneumatic or electronic. A basic mechanism or circuit which is very sensitive is made tractable and useful by a negative feedback element.

In this case the internal feedback is supplied by a pneumatic bellows which controls the position of the upper pivot of the flapper, as in Fig. 3.4. We can make a simple analysis of this mechanism along the following lines. It is assumed that

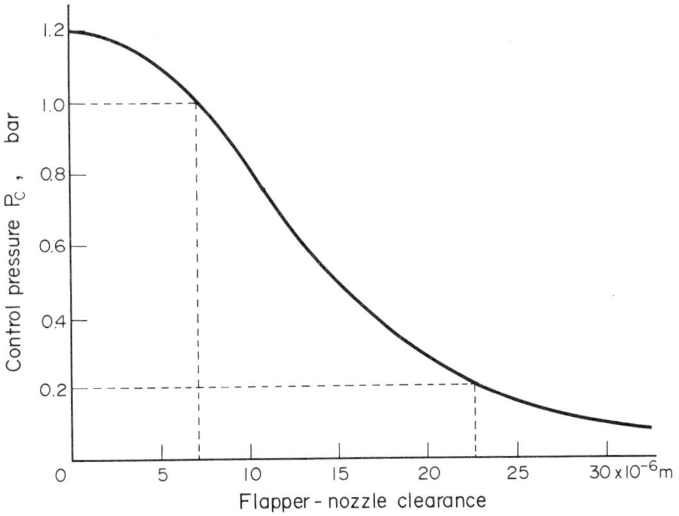

Fig. 3.3. Change in control pressure P_c with flapper-nozzle clearance.

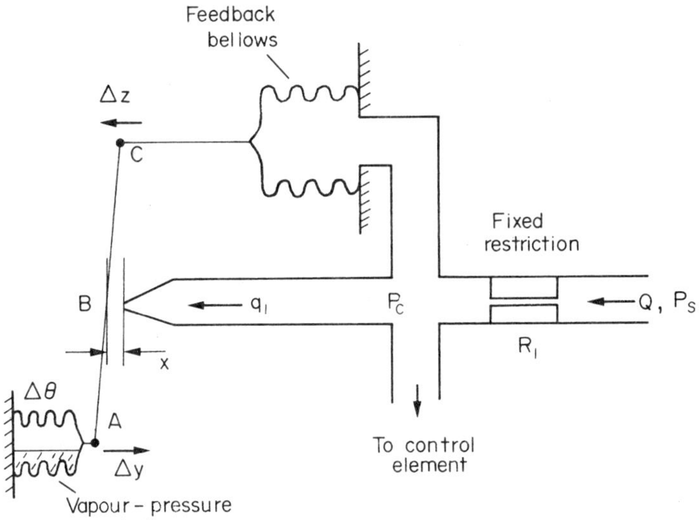

Fig. 3.4. Flapper and nozzle mechanism with feedback bellows.

the flapper is pivoted at both ends, the movements of the ends A and C being designated Δy and Δz respectively. The clearance between flapper and nozzle is x and the change in clearance is Δx.

We define: AC = d,
AB = a,
BC = b,
and a+b = d.

When an increase $\Delta\theta$ occurs in the detected temperature the vapour-pressure bellows will expand by a proportional distance Δy, which is the distance by which the bottom end of the flapper will move forward. This will close down the nozzle and increase the control pressure, causing the feedback bellows to experience a pressure rise and expand by a distance Δz. This will open up the nozzle and there will be a net forward movement which depends on the values of Δy, Δz and the lever-arm ratios b/d and a/d.

The net change in x will be given by

$$\Delta x = \Delta y \cdot b/d - \Delta z \cdot a/d. \tag{3.1}$$

Since Δy is proportional to the change in temperature $\Delta\theta$ and Δz is proportional to the change in control pressure ΔP_c we can write

$$\Delta y = k_1 \, \Delta\theta \quad \text{and} \quad \Delta z = k_2 \Delta P_c$$

where in practice k_1 and k_2 are of the order 10^{-3} m/°C and 10^{-2} m/Pa respectively. Substituting in equation (3.1) gives

$$\Delta x = k_1(b/d)\Delta\theta - k_2(a/d)\Delta P_c. \tag{3.2}$$

Neglecting the flow of air into the feedback bellows, we can write

$$\Delta Q = \Delta q_1 \qquad m^3/s. \tag{3.3}$$

Now it may be assumed that the flow of air through the fixed restriction is laminar, since the diameter is very small, also that the change in flow rate Δq_1 is proportional to the net change in clearance Δx, so that

$$\Delta Q = \Delta P_c/R_1 \quad \text{and} \quad \Delta q_1 = k_3 \Delta x$$

where R_1 and k_3 are of the order 10^8 Pa per m^3/s and 10^5 $m^3/s \cdot m$ respectively. Substituting into equation (3.3) gives

$$\Delta P_c/R_1 = k_3 \Delta x$$

and substituting for Δx from equation (3.2) gives

$$\Delta P_c/R_1 = k_1 k_3 (b/d)\Delta\theta - k_2 k_3 (a/d)\Delta P_c$$

rearranging:

$$\Delta P_c = \left[\frac{k_1 k_3 (b/d)}{1/R_1 + k_2 k_3 (a/d)}\right] \Delta\theta$$

$$= \left[\frac{k_1 k_3 \, b \, R_1}{d + k_2 k_3 \, a \, R_1}\right] \Delta\theta$$

$$\simeq \left[\frac{k_1 k_3 \, b \, R_1}{k_2 k_3 \, a \, R_1}\right] \Delta\theta$$

since $k_2 k_3 \, a \, R_1 >> d$.

Hence $$\Delta P_c = \left[\frac{k_1}{k_2} \cdot \frac{b}{a}\right] \cdot \Delta\theta = K_p \Delta\theta \qquad (3.4)$$

so that the change in control pressure is seen to be proportional to the change in detected temperature, the constant of proportionality being the proportional gain constant K_p. The ratio k_1/k_2 is fixed by the design, but K_p also depends on the ratio b/a, and it is arranged in practice that b/a can be adjusted at will.

The change in control pressure ΔP_c is fed out by the branch air connection to operate a valve or damper actuator to perform the control function.

This type of controller is called continuous-bleed because air is continuously being lost from the nozzle. This forms a load on the compressor which has to make up this loss. It is also a source of noise which may be objectionable, for example in a prestige or highly critical application. This drawback is overcome to some extent by the so-called non-bleed control mechanism, described in the next section.

3.1.2 Non-bleed Proportional Controller

The basic mechanism of this type of controller is shown schematically in Fig. 3.5. There are three ports, one (M) connected to the main supply pressure of 1.2 bar, one (O) exhausting to atmosphere and the branch (B) connection which is connected to the pneumatic actuator or motor. The main and exhaust ports are closed with small valves V_m and V_o which are operated by a lever arrangement. The pushrod CD is connected at its upper end to a temperature sensor which generates a force on the pushrod. Figure 3.5(a) shows the condition when the temperature is such that branch pressure is increasing. As with the flapper and nozzle systems, there is a high gain element which by itself would produce an oscillatory on/off action. The gain is rendered manageable by a feedback element which in this case is the diaphragm EE'. As the air pressure inside the chamber increases, the upward force increases. The pushrod CD is fixed to the diaphragm so it experiences the upward force. The effect is to close the valve V_m. Figure 3.5(b) shows the mechanism in the balanced condition, where the valves are both closed and the branch pressure is held by the diaphragm. The beneficial effect of this feedback diaphragm can be seen by considering the situation which would occur if a leak developed in valve V_o. The pressure would drop and the upward force on the underside of the diaphragm would decrease, with an effect equivalent to that of an increased downward force. The downward force on the plug of V_o would increase and this would attempt to reduce the outward leakage of air, and halt the drop in pressure. Similarly, a leak in the main valve V_m would cause an increase in pressure, tending to close down V_m. In Fig. 3.5(c) the mechanism is seen responding to a reduction in the downward force on the pushrod by opening V_o and allowing air to bleed out to atmosphere. The designation "non-bleed" is not quite correct since air is allowed to bleed out when the branch pressure is reduced. At other times there is no loss of compressed air to the atmosphere and therefore no extra load on the compressor.

The mechanism just described is not capable of control by itself. It needs to have some arrangement for adjustment of set point and proportional band. It therefore has to be incorporated into a complete controller. This mechanism is often called the "relay" although this is perhaps inadvisable because there are many different kinds of relay used in pneumatic control. Figure 3.6 shows how this non-bleed mechanism is used in a proportional controller for temperature control. The temperature sensor feeds a bellows via a capillary tube. The end of the bellows bears on one side of the T-shaped pivoted piece ABC. An adjustable compression spring opposes the thrust of the bellows, and this adjustment allows choice of the set point. The movement of the piece ABC is transmitted by the lozenge-shaped piece D down to a long arm pivoted at one end and whose other end connects

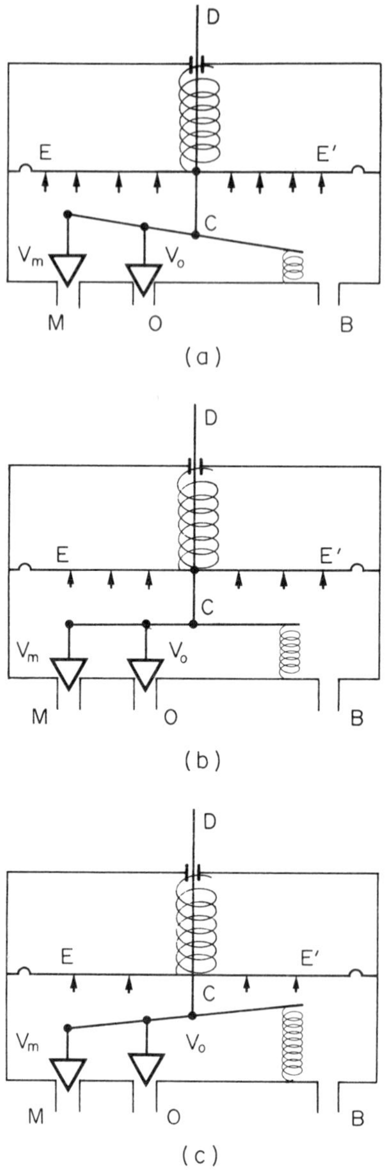

Fig. 3.5. Non-bleed proportional control relay (a) with branch pressure increasing, (b) in balanced condition, (c) with branch pressure decreasing.

to the pushrod of the relay shown in Fig. 3.5. When the lozenge-shaped piece D is on the right of the pivot of ABC then an increase in temperature at the sensor will cause a downward movement of the pushrod and an increase in the branch air pressure from the relay. This is termed "direct acting" (DA). For some purposes it is necessary to arrange that the branch pressure decreases when the sensed temperature rises. This can be brought about by moving D to the other side of the pivot of ABC. The controller is then in a "reverse acting" (RA) mode. Thus the same controller can be used both with those final control elements and

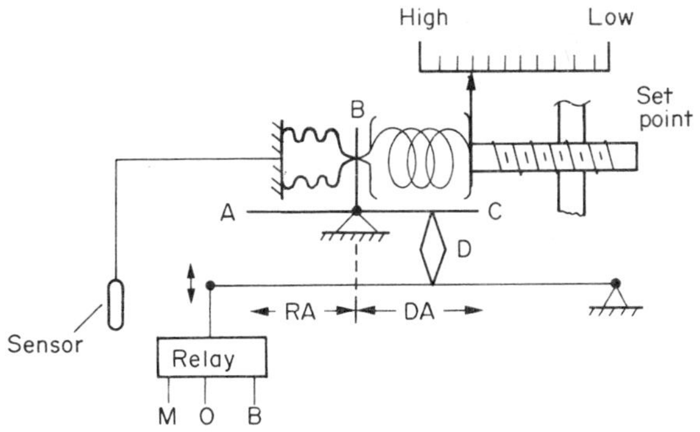

Fig. 3.6. Non-bleed relay used in a proportional controller.

actuators which require "air to open" and those which require "air to close". The proportional band is also adjustable by choice of the position of D in relation to the pivot of ABC. The further from the pivot the piece D is moved, the smaller will be the proportional band, so the bar AC can be calibrated with two scales of proportional band, one for each side of the pivot.

3.1.3 Three-term Pneumatic Controller

The pneumatic controllers described so far have had proportional action only. It is possible to incorporate integral and derivative terms in a number of ways and this section describes the operation of a continuous-bleed type three-term controller. It is an elaboration of the mechanism described in Section 3.1.2, with the addition of an extra feedback bellows, as shown in Fig. 3.7. The tubes connected to these bellows now have adjustable restrictors in them, shown as fluid

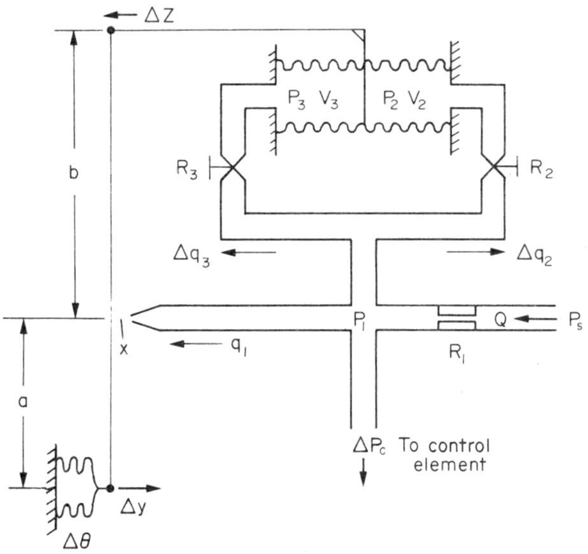

Fig. 3.7. Continuous-bleed three-term controller.

resistances R_2 and R_3. As before, the capsule is considered to experience an increase in temperature $\Delta\theta$, which causes an expansion and a forward movement Δy of the bottom end of the flapper. The change in pressure causes a movement Δz of the top end of the flapper. The net change Δx in the flapper-nozzle clearance is then given by

$$\Delta x = \Delta y\ (b/d) - \Delta z\ (a/d). \tag{3.5}$$

The movement Δy can be assumed to be proportional to $\Delta\theta$. In this case Δz will be proportional to the difference in pressure increase in the two bellows. Substituting in equation (3.5) gives

$$\begin{aligned}\Delta x &= k_1\ (b/d)\Delta\theta - k_2\ (a/d)\cdot(\Delta P_2 - \Delta P_3)\\ &= k_1\ (b/d)\Delta\theta - k_2\Delta P_2\ (a/d) + k_2\Delta P_3\ (a/d).\end{aligned} \tag{3.6}$$

We can apply the ideal gas equation to the air in the right-hand bellows:

$$P_2V_2 = m_2\ RT \tag{3.7}$$

where P_2 = absolute pressure in bellows, Pa,

V_2 = volume of bellows, m^3,

m_2 = mass of air in bellows, kg,

T = absolute temperature, K,

R = universal gas constant, J/kg K.

Differentiating equation (3.7):

$$P_2dV_2 + V_2dP_2 = RT\cdot dm_2 = \rho RT \int \Delta q_2\cdot dt \tag{3.8}$$

where ρ = density of air, kg/m^3,

Δq_2 = flow rate of air, m^3/s.

The first term on the left of equation (3.8) is small compared with the second and can be neglected with little error, so we can write

$$V_2dP_2 \approx \rho RT \int \Delta q_2\cdot dt$$

and differentiating gives

$$V_2\ \Delta\dot{P}_2 \approx \rho RT\ \Delta q_2 = (k_3/R_2)(\Delta P_1 - \Delta P_2)$$

where $\Delta\dot{P}_2 = d(\Delta P_2)/dt$,

$k_3 = \rho RT$,

R_2 = fluid resistance of right-hand restrictor;

rearranging and taking the Laplace transform of the variables gives

$$\Delta P_2(s) = \left[\frac{1}{1 + T_2 s}\right] \Delta P_1(s) \tag{3.9}$$

where $T_2 = R_2V_2/k_3$.

A similar derivation can be obtained for the left-hand bellows, giving

$$\Delta P_3(s) = \left[\frac{1}{1 + T_3 s}\right] \Delta P_1(s) \qquad (3.10)$$

Now the total flow Q from the supply pressure P_s is made up of the flow q_1 through the nozzle, plus the small flows Δq_2 and Δq_3 to or from the bellows. An approximate volume balance will be

$$\Delta Q = \Delta q_1 + \Delta q_2 + \Delta q_3$$

$$\approx \Delta q_1$$

since Δq_2 and Δq_3 are both small.

Hence $\Delta P_1/R_1 = k_4 \Delta x$,

substituting (3.6) into this gives

$$\Delta P_1/R_1 = k_4 \left[k_1(b/d)\Delta\theta - k_2(a/d)\Delta P_2 + k_2(a/d)\Delta P_3\right].$$

Transforming this and substituting from (3.9) and (3.10) gives

$$\Delta P_1(s)/R_1 = k_1 k_4 (b/d)\Delta\theta(s) - \frac{k_2 k_4 (a/d)}{(1 + T_2 s)} \Delta P_1(s) + \frac{k_2 k_4 (a/d)}{(1 + T_3 s)} \Delta P_1(s) \qquad (3.11)$$

rearranging:

$$\frac{\Delta P_1(s)}{R_1} + \frac{k_2 k_4 a}{d}\left[\frac{1}{1 + T_2 s} - \frac{1}{1 + T_3 s}\right] \Delta P_1(s) = \frac{k_1 k_4 b}{d} \Delta\theta(s).$$

The first term on the left-hand side may be neglected since R_1 is very large, which gives

$$(k_2 a/k_1 b)\left[\frac{1}{1 + T_2 s} - \frac{1}{1 + T_3 s}\right] \Delta P_1(s) = \Delta\theta(s)$$

and

$$(k_2 a/k_1 b)\left[\frac{(T_3 - T_2)s}{(1 + T_2 s)(1 + T_3 s)}\right] \Delta P_1(s) = \Delta\theta(s),$$

and finally,

$$\Delta P_1(s) = Kp\left[1/T_i s + 1 + T_d s\right] \Delta\theta(s) \qquad (3.12)$$

where

$$Kp = \frac{k_1\ b\ (T_3 + T_2)}{k_2\ a\ (T_3 - T_2)},$$

$$T_i = T_3 + T_2,$$

$$T_d = T_2 T_3/(T_2 + T_3).$$

Equation (3.12) is of the same form as that required for a three-term controller. Unfortunately in this case the proportional gain and the integral and derivative action times are not simply related to the two time constants T_2 and T_3. The outflow of air from the controller to the control element or actuator was not included in the volume balance because the controller is normally connected to a relay which is able to supply a large volume of air at the required control pressure P_c.

3.2 ELECTRIC CONTROLLERS

Electric controllers using no electronic amplifiers are rather rare nowadays: they may be said to be obsolete. A brief description is given here for completeness, and because they may still be seen in existing installations.

The central component of these systems is shown in Fig. 3.8. The balanced relay has two separate windings which are electrically isolated from each other, and which carry currents i_1 and i_2. The armature is arranged so as to be able to pivot either clockwise or anticlockwise, carrying the central contact. When the two currents are equal they each exert the same torque on the armature, but in opposite directions so that it does not rotate. This is the balanced state, and the terminal A is not connected to the other terminals B and C. If now the current i_1 rises and i_2 remains the same, the armature will rotate anticlockwise and contact will be made between terminals A and C. Conversely, when i_2 is slightly greater than i_1 the armature will rotate clockwise making contact between terminals A and B.

Figure 3.9 shows the relay connected into a simple operational circuit. The two terminals B and C are connected to the ends of the field windings of a reversible capacitor start motor, which operates on a 24-V a.c. supply from a transformer. The transformer also provides power for the two currents i_1 and i_2. The supply is fed to the circuit via the two sliders at top and bottom. The current i_1 is determined by the sum of the two resistances R_2 and R_4. The current i_2 is determined by R_1 and R_3. When $R_2 + R_4 = R_1 + R_3$ the two currents are equal and the relay is balanced. The motor is not then energised and the valve remains in its current position. When a temperature rise for instance then occurs at the sensor the top slider will move from right to left, causing R_1 to decrease and R_2

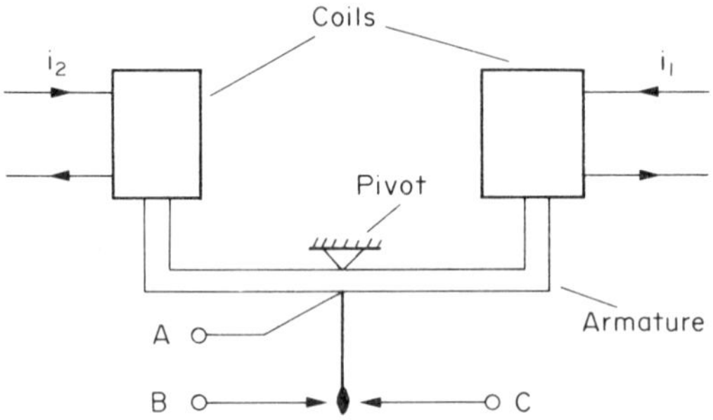

Fig. 3.8. The balanced relay.

to increase. The result will be an increase in i_2 and a decrease in i_1, causing the relay to operate and make contact to terminal B. The left-hand winding of the motor will then be energised and the motor will start. This has two effects: it must turn the valve in such a direction as to reduce the original rise in temperature also it must operate the bottom slider and move it towards the right,

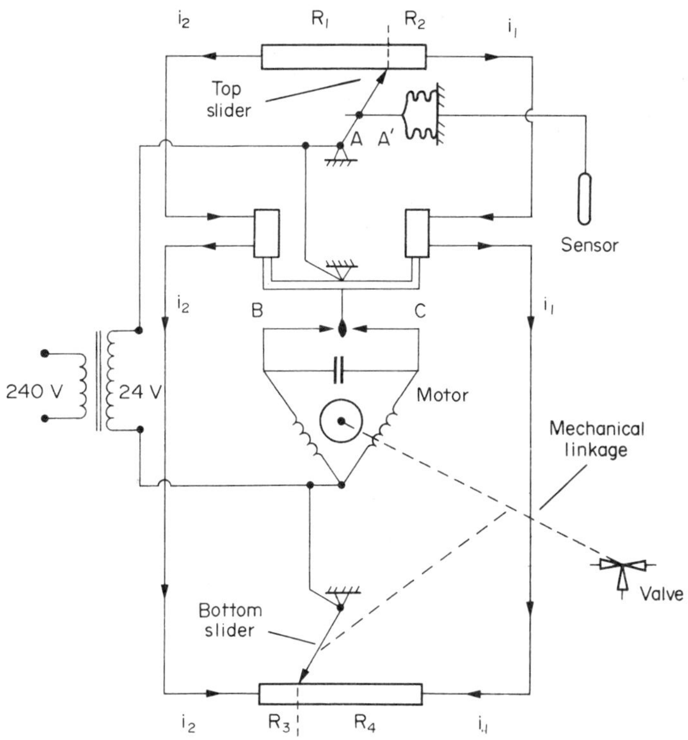

Fig. 3.9. Balanced relay in circuit with a sensor and valve operated by a motor.

increasing R_3, decreasing R_4 and bringing the two currents back to equality.

The action of this controller is proportional. The proportional band can be adjusted by moving the position of the pivot at A up or down the top slider to alter the lever arm ratio. The set point can be adjusted by altering the distance AA'. As the sensed temperature varies over the proportional band range the top slider traverses the length of the top resistor, the motor causes the valve to traverse the full range of its travel from open to closed and the bottom slider also moves over the length of the bottom resistor. The motor itself is a small unit, usually rotating at 1500 r.p.m. In order to make it suitable for valve or damper operation, a reduction gearbox is used giving a final drive shaft speed of about 1 r.p.m. and a corresponding increase in available torque. In order to ensure that the motor really does stop when the valve has reached the ends of its stroke, limit switches are incorporated, as indicated in Fig. 3.10 (LS1 and LS2). Some other details are also included in this Figure, such as the hold-in coils in the relay (H1 and H2). The purpose of these is to ensure that the motor is not started and stopped very frequently, which would produce rapid wear. When the imbalance of currents in the main coils of the relay, M1 and M2, is great enough to cause the relay to operate, the current to the motor field winding also passes through a very small coil (H1 or H2) wound on top of the main relay coil. The relay thus holds itself on until a slight imbalance of currents in the opposite sense pulls the central contact back to its middle position. The effect is analogous to that of hysteresis or differential gap in a thermostat and has the same function: to reduce the frequency of operation.

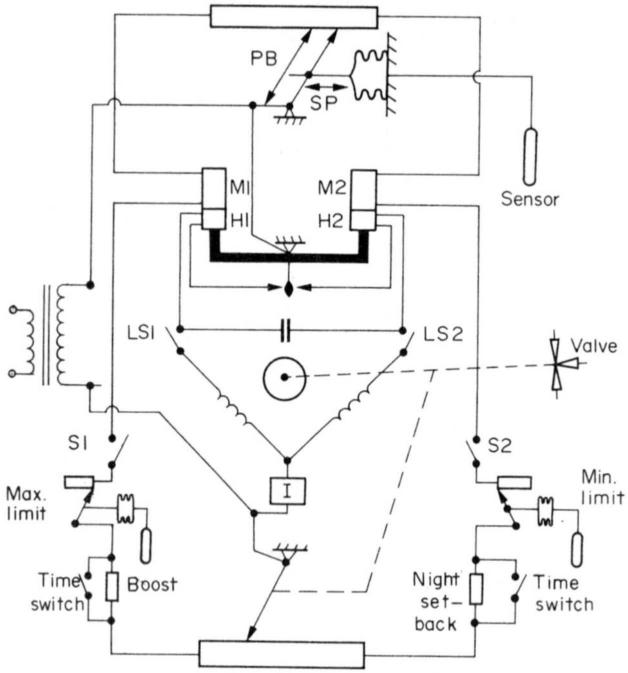

Fig. 3.10. Balanced relay circuit with various additional functions: limit switches, manual control switches, maximum and minimum temperature limits, interruptor and hold-in coils.

The mechanical linkage from the motor gear box which operates the valve is arranged also to operate the limit switches LS1 or LS2 at the two ends of travel of the valve or damper. In some circumstances, e.g. for maintenance purposes, it is desirable to be able to force the valve or damper to travel to its fully closed or fully open position. This can be done by opening one or other of the manual switches S1 or S2. In order to set limits on the maximum or minimum temperature of the supply medium, the circuit can be arranged with extra sensors as shown. These operate sliders which introduce extra resistance into the appropriate side of the circuit when the sensed temperature goes above or below the set limit. This can be desirable in heated air systems to avoid either chilling the occupants with a cold draught or burning them with overheated air. In some heating systems operated intermittently, it is desirable to have a night set-back facility, so that the system runs at reduced temperature overnight. This is achieved by having a small fixed resistor in circuit which is normally shorted out by a switch. When the switch is opened at a predetermined time the current in that side of the circuit is reduced and the motor is forced to drive the heating valve towards its closed position to balance the currents again. Conversely, in order to bring a building up to temperature as quickly as possible in the morning, a timed boost facility can be provided. This consists of the same circuit arrangement as for the night set-back, but it is installed in the other side of the circuit (see Fig. 3.10). Finally, it is sometimes useful to be able to adjust the speed of operation of the valve or damper, where the demand side capacity is small, such as in a non-storage-type heat exchanger. This can be achieved by using an interruptor (marked I in Fig. 3.10) which consists of a motor-driven rotating cam which causes contacts to close for a set time per revolution. The contacts allow current to pass to the motor, which moves in a series of small steps so that the time taken to move over its range is increased.

3.3 ELECTRONIC CONTROL

Electronic control, in contrast to electric control, has circuits which include valves or transistors to give amplification of small voltages from Wheatstone bridges or thermocouples. The usual circuit for control purposes is the Wheatstone bridge and we shall see how it can act as an error detector. The sensors used for pneumatic and electric controllers have to produce a physical motion. This means that expansion or vapour-pressure types have to be used for temperature sensors and such materials as hair or nylon for humidity sensors. Electronic controllers, because of the requirements of the Wheatstone bridge, require sensor types whose electrical resistance value varies with the measured quantity. Platinum or nickel metallic detectors or thermistors can be used for temperature, and the various sensors based on lithium chloride can be used for humidity. This allows much greater flexibility on installation with electronic control than with the other types. At the other end of the control chain the position is rather different, and pneumatic operation has the advantage in speed, flexibility and power over electric motor actuators. Pneumatic actuators are also cheaper than electric. Some systems attempt to combine the advantages of both methods by using electronic detection and control action and pneumatic actuation of the final control element, via electro-pneumatic relays.

3.3.1 The Wheatstone Bridge as an Error Detector

It was shown in Section 1.1.3 how the Wheatstone bridge acts as a detector of change in resistance, and produces an output voltage dependent on the size and the direction of resistance change. In electronic control circuits it is usual to provide alternating supply voltages to the bridge. This is usually at a level of about 10-15 volts and is conveniently available from a mains step-down transformer. This also simplifies the selection of subsequent amplifier units. Figure 3.11 illustrates how the bridge can act as a phase-sensitive error detector in a control circuit. The bridge in Fig. 3.11(a) has a supply voltage V_s, three fixed resistors and one variable resistor R_1. In Fig. 3.11(b), when $R_1 = 100\ \Omega$ the bridge is balanced and the output voltage V_{AB} is zero. When R_1 gets smaller, say to 90 Ω, the voltage drop across R_1 will be smaller than that across R_3 so the bridge will go out of balance and V_{AB} will be in phase with V_s. Conversely, when R_1 rises above 100 Ω, say to 110 Ω, the voltage drop across R_1 will be greater than that across R_3 and the bridge will be unbalanced in the other direction. The phase of V_{AB} will therefore be reversed with reference to V_s.

Hence it appears that the output voltage of the bridge with a.c. supply can indicate not only how much temperature change has occurred, but also in which direction, up or down. This is important in control terms because it is necessary to arrange for the controller to drive the valve or damper in the direction which will oppose the change in controlled variable.

3.3.2 Electronic Proportional Controller

Having established the operation of the bridge circuit as a detector, it can now be incorporated into a control circuit. This is shown in its simplest form in Fig. 3.12. The resistor R_1 is now designated $R(\theta)$ since it is intended to be temperature sensitive. The output voltage of the bridge is connected to a phase-sensitive amplifier (PSA) and the alternating supply voltage V_s is also connected to the amplifier so that comparison can be made of the phases of the two voltage waves V_s and V_{AB}. Depending on the magnitude and relative phase of the bridge output voltage, the amplifier is able to energise one or other of the two relays R_1 or R_2, causing

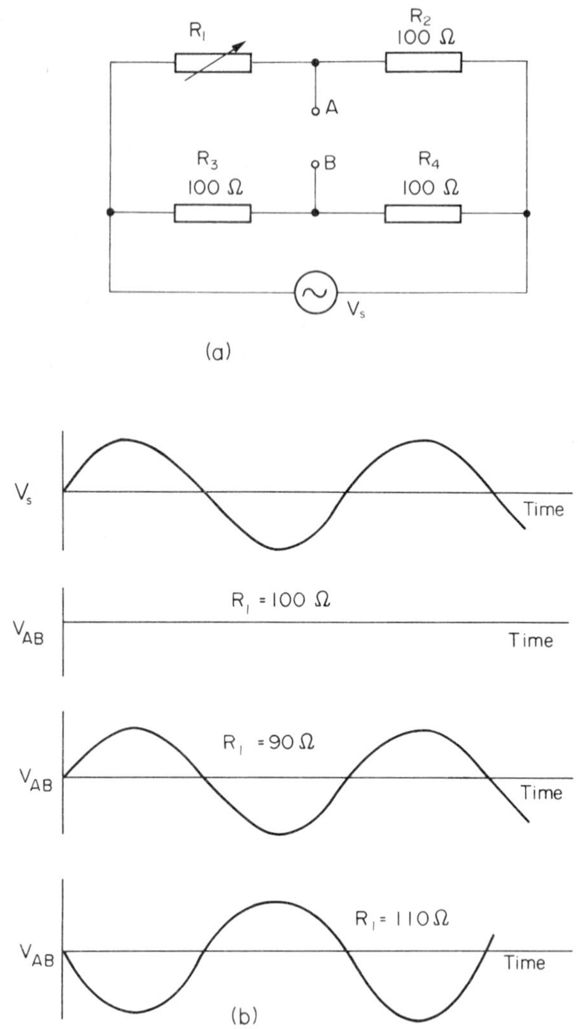

Fig. 3.11. The Wheatstone bridge as a phase-sensitive error detector.

the electric motor to turn in one direction or the other. A bridge unbalance in one direction will thus cause a valve or damper to open: unbalance in the other direction will cause it to close.

In fact the amplifier is normally so sensitive that any very tiny unbalance in the bridge will cause one or other relay to be energised, perhaps momentarily, so that the quality of control would not be very good. As with the other controllers we have seen, the basic system is of very high sensitivity and needs the addition of internal negative feedback to make it useful.

The controller of Fig. 3.12 would be an on/off system with zero hysteresis. The modifications to produce a usable proportional control action are shown in Fig. 3.13. The mechanical linkage between the motor and valve also operates a slider

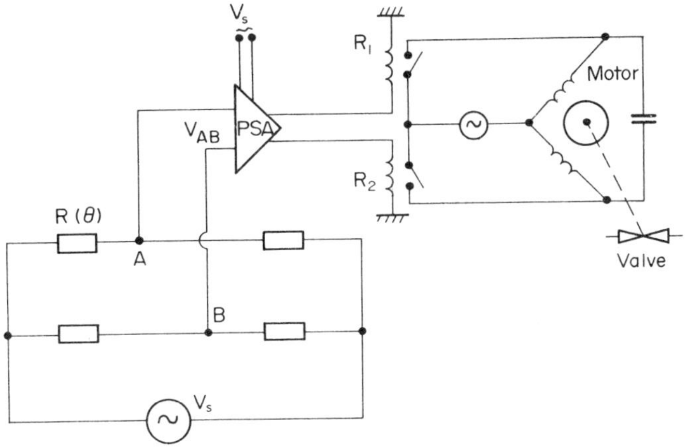

Fig. 3.12. Basic control circuit with Wheatstone bridge.

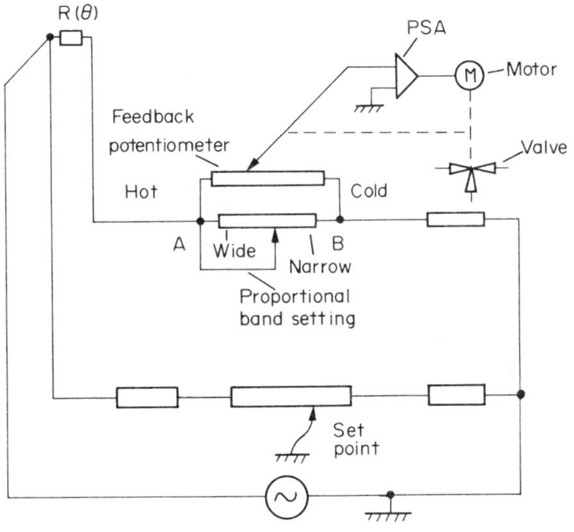

Fig. 3.13. Electronic proportional controller with set point and proportional band adjustments.

moving over a so-called feedback potentiometer. The slider picks-off the voltage output of the bridge and carries it to the amplifier input. The phase-sensitive amplifier and its relays for operating the valve are the same as in Fig. 3.12. It should be noted that the negative input terminal of the amplifier, one side of the bridge supply and the set-point potentiometer slider are all connected together in common. The mechanical linkage from the motor is arranged so that the slider always moves so as to seek a point at the same potential as this common point, called the neutral point. If it is assumed that the system is at rest at a certain temperature, equal to the set point, and then the sensed temperature θ rises, then the neutral point will move away from the slider on the feedback potentiometer
For a metallic detector the resistance R(θ) will increase when the temperature rises and the neutral point will move to the left. The input voltage to the amplifier

will now no longer be zero, so the motor will be started, causing the valve to close down slightly (assuming a heating duty) and the slider to be moved slightly to the left, following the neutral point. If the temperature θ continues to rise, the valve will continue to close down and the slider continue to move to the left until it reaches the end of the feedback potentiometer and a limit switch will operate. To make the slider traverse the whole length of the feedback potentiometer, the value of $R(\theta)$ must change by twice the resistance between the points A and B. Obviously this corresponds to the proportional band of the controller. Adjusting the resistance between A and B gives a means of adjusting the proportional band. The resistance of the feedback potentiometer itself is fixed, usually at 135 ohms. The overall resistance can be adjusted by connecting a second potentiometer in parallel with the feedback potentiometer, arranged as shown in Fig. 3.13, so that its overall resistance can be adjusted. This arrangement does not interfere with the movement of the slider over the feedback potentiometer, which still traverses the whole length. The proportional band setting potentiometer can be a front panel control allowing the operator to set up the desired value of proportional band. Referring to Fig. 4.3, the offset will be zero when the valve is at 50% of its stroke, and in this condition the slider of the feedback potentiometer is exactly at the centre of its travel. The temperature at which this occurs can be chosen by altering the ratio of resistances in the two bottom arms of the bridge. This can be done with the set-point adjustment as shown in Fig. 3.13. Moving the slider to the left reduces the set point when a metallic detector is used, and increases it when a thermistor is used.

The various additional functions which were mentioned for the balanced relay circuit in Fig. 3.10 are also available with the Wheatstone bridge circuit. These are indicated in Fig. 3.14.

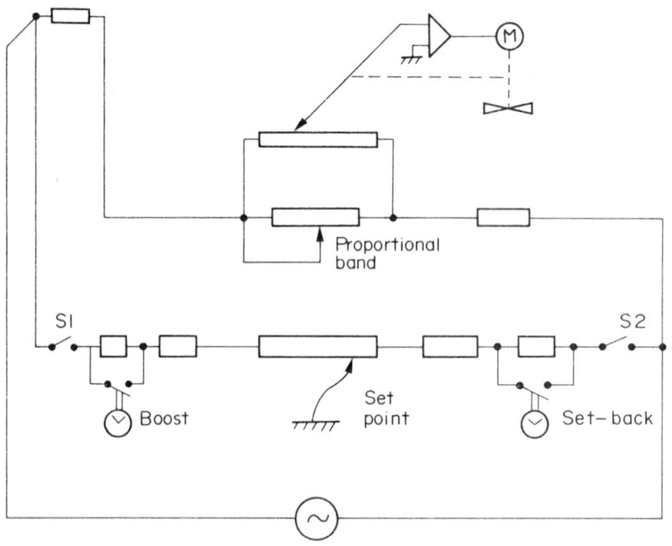

Fig. 3.14. Electronic proportional controller with boost and set-back functions controlled by a clock. Switches S1 and S2 are manual.

This type of controller can also be arranged to use more than one bridge circuit or an extended bridge circuit for special purposes. With an outside compensator-control arrangement, the temperature at which the flow water from a heating boiler is controlled may be dependent on the outside temperature. Obviously, when the outside temperature is low, a higher flow water temperature will be needed to maintain the same inside temperature. The type of relationship between the two temperatures is indicated in Fig. 3.15. A change of about 20°C in outside temperature must produce a change of about 40°C (90-50) in flow-water temperature and θ_F must increase as θ_o decreases. This is a simple and convenient way of providing overall control on a building heating system without the necessity of installing individual room-temperature detectors. Because the slope of the line in Fig. 3.15 is -2, the two variables θ_o and θ_F must be arranged to have different

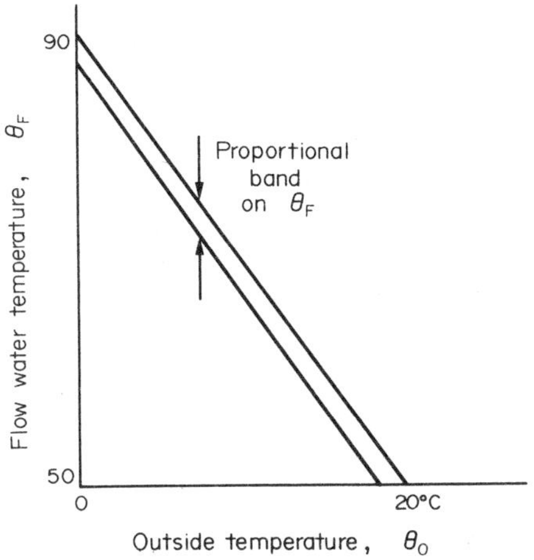

Fig. 3.15. Outside compensator - typical relationship between θ_F and θ_o

power of control over the valve position, although the direction of control must be the same for both variables. An increase in θ_F must cause an opposing change in valve stroke, so as to bring θ_F back to its previous value; an increase in θ_o must also cause the valve to reduce θ_F in accordance with Fig. 3.15. This difference in power of control by two sensors over the same controlled variable is called authority; in this case the authorities must be in the ratio of about 2:1. Figure 3.16 shows two methods of providing different authorities. In Fig. 3.16(a) the two sensors are placed in the same bridge circuit, diagonally opposite arms having the same effect. The different authorities are produced by putting adjustable resistances in parallel with the sensors. Figure 3.16(b) shows the two sensors connected in separate bridge circuits which are identical except for their supply voltages. The two bridges are connected in series and their output voltages add algebraically.

The heat-loss rate of a building with a given inside temperature is dependent not only on the dry-bulb temperature but also on the solar radiation intensity and on the wind speed. To take these into account, more complex circuits have been developed, as shown in Fig. 3.17. The additional two bridges produce output voltages which are dependent on the wind and sun effects and cause the curve shown in Fig. 3.15 to be displaced depending on the instantaneous values of these variables. The various sensors are built into a casing which has to be located on an outside surface of the building. For simple outside temperature detectors this is fairly straightforward and it is usually recommended that the detector be placed on a north-facing wall, so that it is shaded from solar radiation effects. For so-called "weather compensators" which respond to solar and wind effects, the siting of the detector and adjustment of the authority of the different sensors can be a

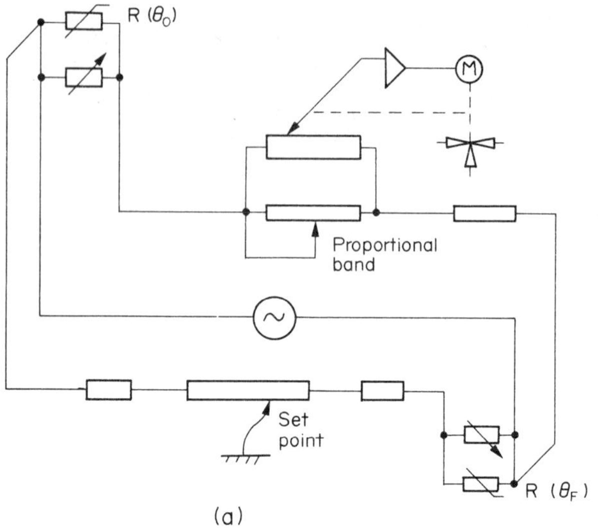

Fig. 3.16(a). Authority of sensors adjusted by choice of sensor resistance.

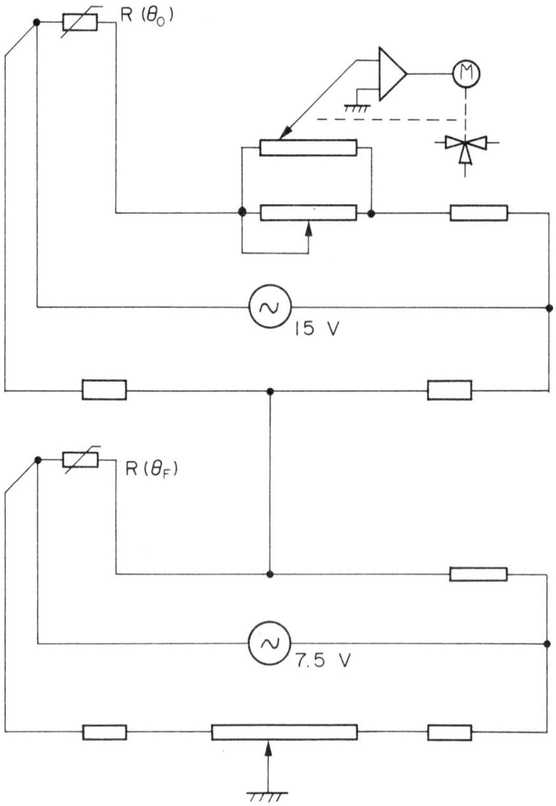

Fig. 3.16(b). Authority of sensors adjusted by bridge supply voltage.

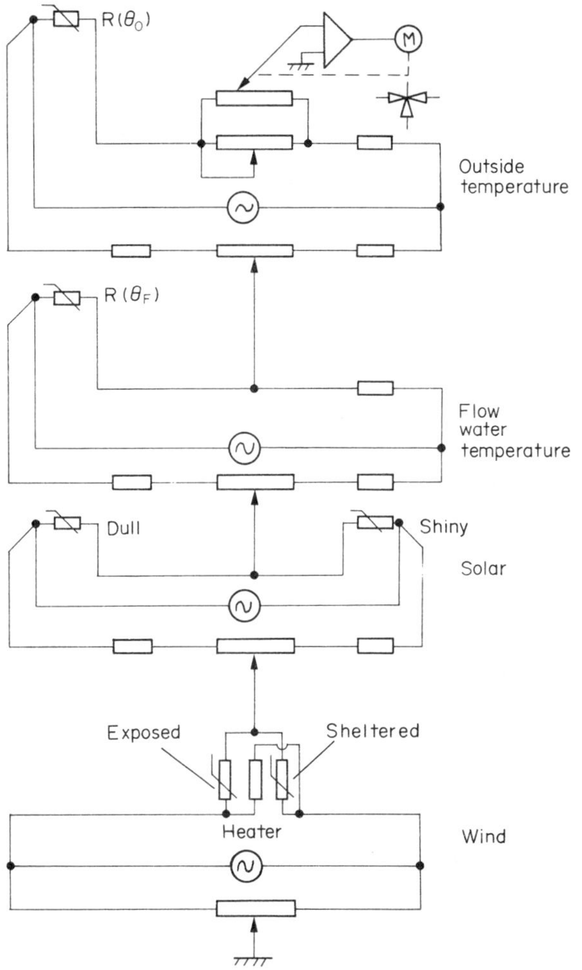

Fig. 3.17. Weather compensator circuit.

matter of some difficulty. It is prudent to mount the detector on that face of the building which is likely to suffer the largest heat loss, although this carries the danger of overheating in rooms located on other faces. Subdivision of larger buildings into zones, each with its own compensator system, can be successful. The outside compensator principle, in any case, cannot provide accurate control of room conditions since it is open-loop. It is probably best regarded as a rather coarse control, with finer secondary control being provided where necessary by individual closed-loop controllers.

As an alternative to having one valve influenced by a number of sensors simultaneously, it is possible to arrange for one sensor to control the position of a number of valves or dampers in sequence. An example of this sequence control is when an air-temperature sensor is installed in a supply duct after a spray washer so that it is sensing the temperature of saturated air. The temperature control is then equivalent to control of moisture content. The layout of plant is shown in Fig. 3.18(a) and the corresponding bridge circuit in Fig. 3.18(b). A further heater

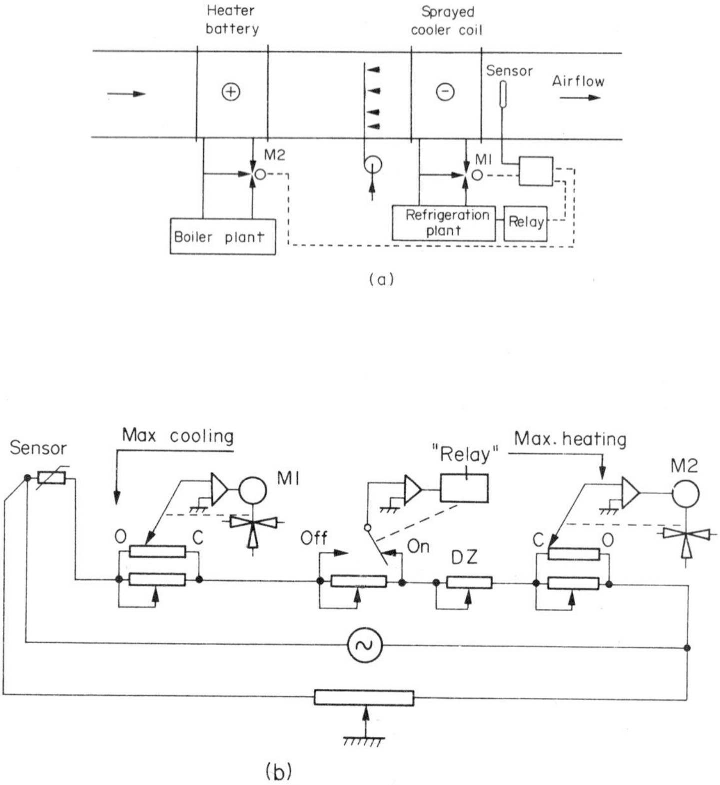

Fig. 3.18. Sequence control: (a) air-conditioning plant to give dewpoint control: (b) bridge circuit for sequence control of valves, with dead zone (DZ) between heating and cooling.

battery with its own controller can then provide supply air of the desired dry-bulb temperature. Obviously this same principle of sequencing a number of final control elements can be applied to a number of other cases such as face and bypass dampers or summer/winter applications.

3.4 SWITCHING THERMOSTATS AND HUMIDISTATS

These devices are able to operate electrical contacts or switches when the sensed variable, temperature or humidity, exceeds or falls below some preset level. The sensing element therefore has to be one which produces a mechanical movement or force in response to a change in the sensed condition. For temperature, the suitable sensors are bimetallic strips or coils, rod and tube elements, vapour-pressure capsules or bellows or capillary-tube versions of the same type (see Section 1.1.1).

The temperature or humidity around the sensor may often change only slowly. If a simple contact is fixed to a bimetal strip to produce the control action, as shown in Fig. 3.19, the contact closure and opening will occur slowly and there will be long periods of time when the contact gap will be very small. This will produce

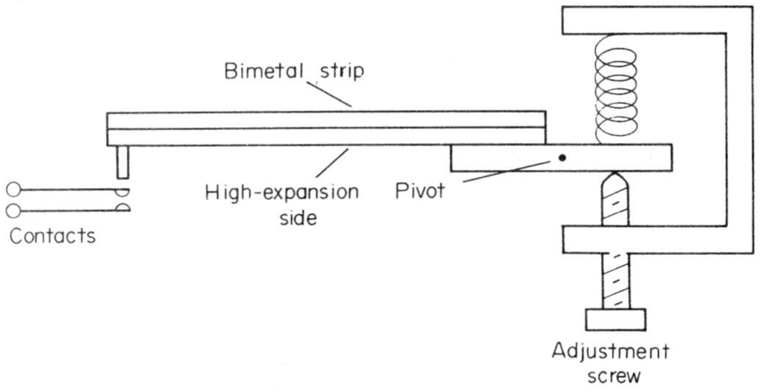

Fig. 3.19. Simple mechanism for thermostat - no snap action.

arcing at the contacts and poor contact pressure which causes heat generation at the contacts. The control would be unreliable and the life of contacts would be short. In order to avoid this, a positive make-and-break action is introduced, usually called snap action. This can be produced using a small permanent magnet or a spring with a "top-dead-centre" type of effect, as shown in Fig. 3.20. The contacts cannot open until the stress in the bimetal is sufficiently great to

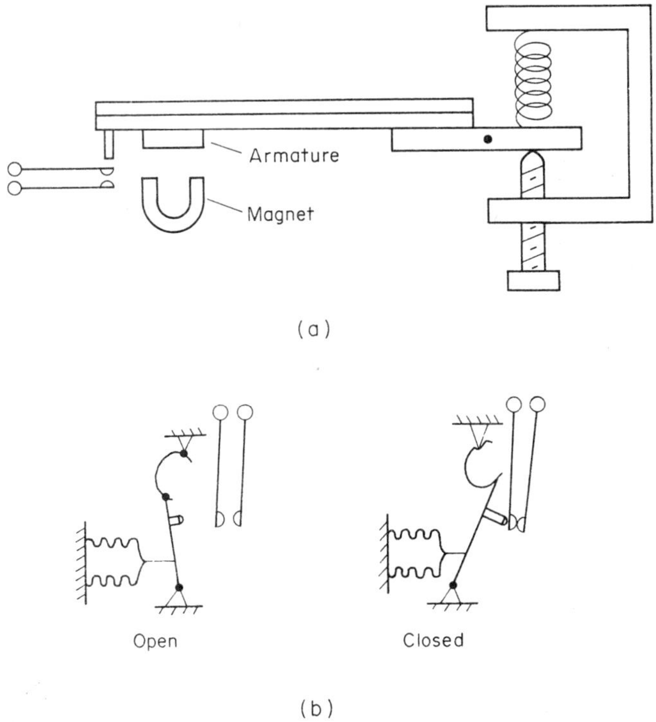

Fig. 3.20. Mechanisms for thermostats - with snap action, (a) using a magnet, (b) using a spring.

overcome the magnetic force between the magnet and armature or the spring force. Similarly on closure, a force has to be generated to produce contact. The make and break actions are now very rapid, and very little wear of the contact surfaces should occur, especially with resistive loads. With inductive loads, the current rating of the contacts may have to be reduced. With 50 Hz a.c. mains, the time taken by one cycle is 20 milliseconds and the instant of contact opening or closure may be at any point in the cycle. If a closure happens to occur near a voltage maximum, the current drawn from the mains supply will rise very rapidly, with sinusoidal components which extend well into the radio-frequency region. This produces radio-frequency interference (RFI) which appears as annoying clicks and buzzes on television and radio sets when switching occurs. In order to avoid this, it is possible to use one of a number of integrated circuits which are now available and which also avoid the necessity of mechanical contacts. The integrated circuit monitors the voltage waveform of the mains supply and at the instant when the voltage passes through zero it triggers a thyristor, so that switch-on of the current is achieved without surges due to switching on when a non-zero voltage exists. These integrated circuits are known as "zero-crossing detectors". They are not yet built in to simple room thermostats because of expense, but will probably become more widely used as the price of solid-state switching components falls.

Returning to the simple type of thermostat mechanism, it might appear that one such as is shown in Fig. 3.20 would give satisfactory operation. In fact the swings in temperature resulting from control with such an arrangement would be rather wide and the control would probably not be satisfactory. The reason for this is the lack of any internal feedback. If the application to a heated room is considered, when the thermostat initially calls for heat and switches on an electric heater, the contacts will remain closed while the heater itself warms up and gives off heat to the air of the room. This warmed air has to travel across to the thermostat itself and raise its temperature before the contacts will open. Meanwhile, heat stored in the mass of the heater will be emitted and cause overshoot of temperature. Similarly, when the temperature falls, it takes time for the corrective action of the thermostat to take effect, and a large drop can be produced. The on/off cycle will be relatively slow, perhaps taking up to 1 hour for a complete cycle.

This effect of wide swings and slow cycling can be improved by introducing an internal feedback. A small resistor is installed in the casing of the thermostat, close to the temperature sensor. It is arranged that, when the contacts are closed and the main heater is on, a current passes through this small resistor, generating a small amount of heat and causing the thermostat to switch off more rapidly than it would otherwise. For this reason the small heater is called an accelerator. It introduces a "minor loop" of feedback within the main loop, as shown in Fig. 3.21. The accelerator heaters used in room thermostats may be arranged in two alternative ways; either as a rather low resistance (less than 1 ohm) in series with the load or as a high-value resistance (about 0.1 Mohm) in parallel with the load. The series type are often used for 24-volt operation, and are usually adjustable to permit allowance for a range of load currents. The parallel arrangement is used for 240-volt operation; it is probably more reliable than the series type, and for a given supply voltage it needs no adjustment.

The effect of the accelerator is to produce a rapid heating or cooling effect to the sensor when the main heater is switched on or off. Instead of having to wait for the heat from the main room heater to reach the thermostat and switch it off, the thermostat is cut off relatively early and cools relatively quickly, with the contacts closing again in a short time. In effect the heat is released in short bursts rather than large chunks, giving more rapid switching and a smaller amplitude of oscillation. This accelerator heater is generally sized to provide a temperature rise within the casing which equals the total hysteresis band of the switch, about $2^{o}C$.

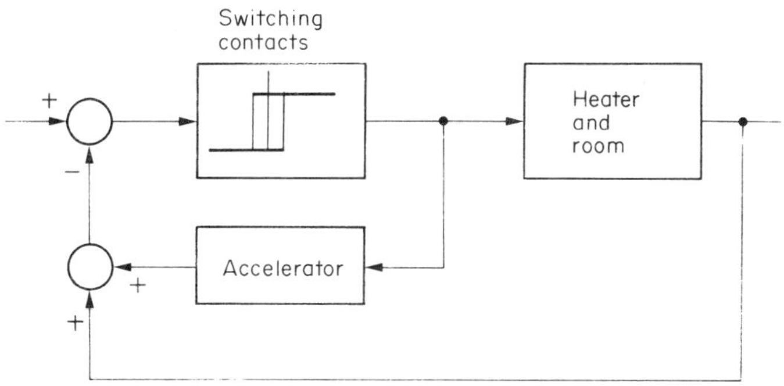

Fig. 3.21. Accelerator heater generating a feedback loop within the main feedback path.

It is not always understood that the shape of the cycle of controlled temperature depends on the load, i.e. on the difference between inside and outside temperature. The change in load also affects the proportion of "on" time and hence the mean temperature rise of the thermostat sensor above room temperature. These effects combine to produce a crude form of proportional control action from the accelerated thermostat. The effect of various load levels is shown in Fig. 3.22. When the

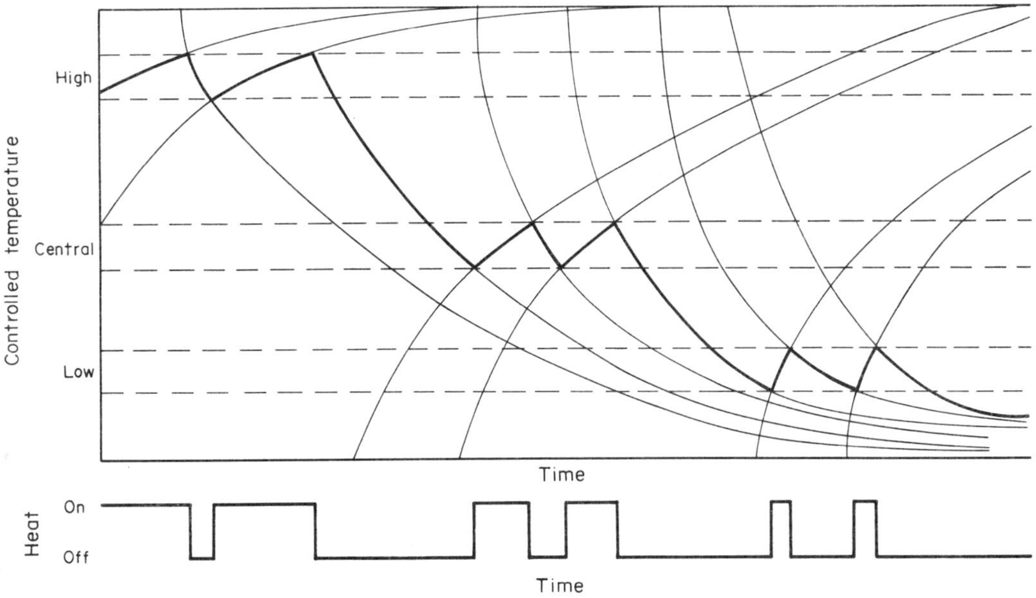

Fig. 3.22. Temperature cycle profiles for different load conditions.

load is high, i.e. when the set point of the thermostat is high, the mean temperature over the cycle is lower than the set point. When the load is low, with a low set point, the mean temperature over the cycle is higher than the set point. When the set point is at the centre of the control range, the two coincide. This central position also gives a minimum cycle time and amplitude. A typical set of characteristics is shown in Fig. 3.23(a), (b) and (c), for a thermostat with 2°C

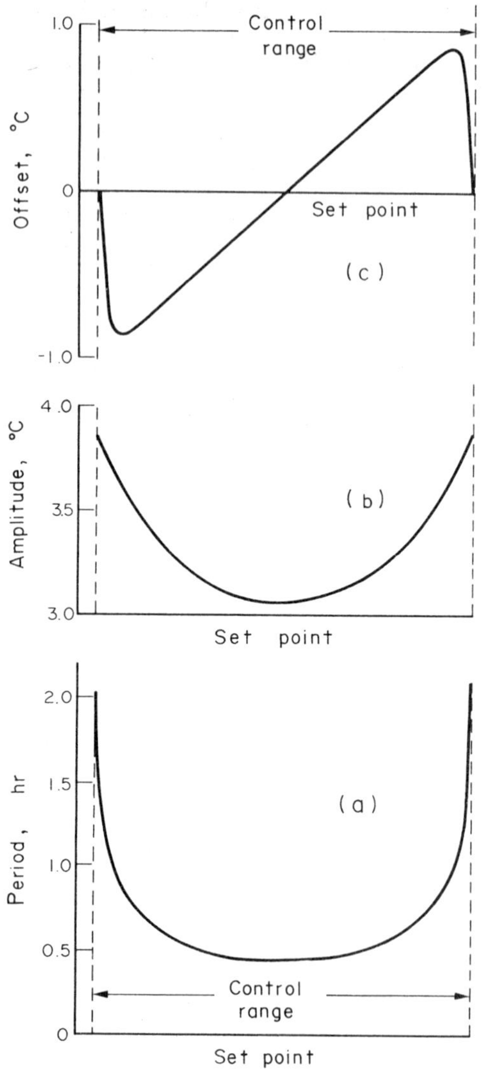

Fig. 3.23. Characteristics of controlled cycle for range of loads. (a) cycle period, (b) peak-to-peak amplitude, (c) offset = set point-mean controlled temperature.

total hysteresis. The control range is the change in temperature produced by putting the heater on continuously. When the set point is in the centre of the control range, both the period and amplitude of the controlled cycle are at a minimum, and the offset if zero. Moving the set point away from the centre of the control range causes both the period and the amplitude to increase and the offset becomes non-zero.

In order to reduce this offset effect, it is possible to incorporate a second small heater, known as a compensator, in the thermostat casing. The accelerator heater is energised when the room heater is on, but the compensator is energised when the room heater is off. Instead of being clipped to the sensor and influencing it immediately, the compensator is installed within the base of the thermostat, and its heating effect is smoothed out. The result is to give an action similar to that of an integral control term, to remove the offset between set point and mean room temperature.

In addition to these heaters, it is possible to obtain a thermostat with an extra, auxiliary, heater in the casing. This may be controlled by a time switch which can be integral with the thermostat or may be a separate unit. Its effect is to raise the temperature of the sensor by about 5°C so that the room temperature is then maintained below the value set on the scale. Its purpose is to act as a "set-back" device to economise on fuel by reducing the heat loss during the night or over a week-end. A manual switch is usually provided to override the time-switch when desired.

In order to obtain the best results with individual control by room thermostats, care must be exercised in their location in the room. The aim should be to site the thermostat sensor in a place which is representative of average conditions in the room. They should be:

about 1.5 m above floor level,
not in the direct rays of the sun,
on an internal wall,
not over a heat emitter,
not near a window or door,
not exposed to draughts.

3.5 STEP CONTROLLERS

Ordinary room thermostats are two-position switch devices, capable of switching on or off a single device. It is sometimes necessary to switch a number of devices in sequence, such as sections of a heater battery or stages of a refrigeration compressor. In such cases it is convenient to use a set of cam-operated switches, operated in sequence by a set of cams on a common shaft. These usually take the form of microswitches and there may be up to twelve in a single step controller. Each microswitch should have a changeover action so that a make or a break operation can be used. Where the current rating of each switch is insufficient for the duty required, they can be used to energise a set of contactors or relays of suitable rating. This is illustrated in Fig. 3.24, which shows a step controller used to switch successive stages in an electric heater battery. The camshaft is arranged to operate the switches R, 1, 2, etc., in sequence as it rotates. If a number of the switches are operated and at some instant the power supply is switched off or a power failure occurs, a large current surge will occur when power is switched on again. This is undesirable and could cause circuit-breakers to trip. To avoid this, a recycling relay is incorporated, which forces the step controller to run back to the starting position (no switches operated) when power is first applied. Figure 3.24 shows the system in the non-energised condition. Because switch A in the recycling relay is open, the Wheatstone bridge circuit of the controller is thrown out of balance and forces the motor to drive back and open the switches in the sequence ..., 4, 3, 2, 1, R. When the switch labelled R is opened (the centre contact moving to the left) the coil of the recycling relay is energised, provided that the simple fan interlock relay is energised. This causes switches A, B and C to cross over to the energised condition (centre contact to the right). The Wheatstone bridge circuit is now made through switch A and the sensor takes over control. Switch B provides power available to the coils of the contactors and switch C holds the recycling relay on. If the sensor detects a fall in temperature and calls for heat, the motor will rotate the camshaft so as to operate the switches R, 1, 2, etc., in sequence (centre contact moving to the right). When R operates the recycling relay still

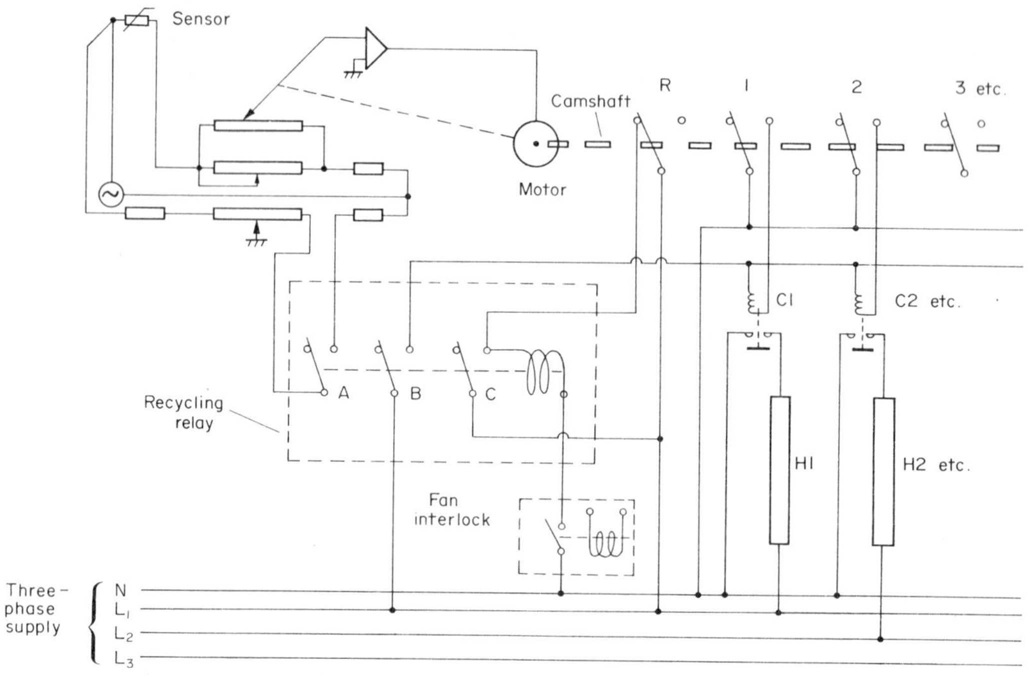

Fig. 3.24. Step-controller circuit with recycling relay and fan interlock.

remains energised via contact C. As each switch 1, 2, 3, etc., operates it is able to energise one of the contactors C1, C2, etc., to switch on successive stages of the heater battery.

Electronic types of step controller can be obtained where the setting-up is done by setting voltages instead of by adjusting cams. With these types it is possible to reduce the number of stages but have them of different ratings, increasing by a factor of 2. With n stages it is then possible to have 2^n steps in the output. For example, an output from 0 to 15 kW can be obtained by combinations of four stages of 1, 2, 4 and 8 kW.

Pneumatic units with the same function as ordinary step controllers are also available. A number of capsules are fed by a controlled air pressure and are arranged so that each operates an electrical switch at some predetermined air pressure.

3.6 TIME SWITCHES

Most building services plant is operated intermittently to some extent: it is sometimes said that hospitals and prisons are the only buildings which require continuous running of plant every day of the year. Switching on and off can obviously be done manually, and provision is always made for this to be done, but it is more convenient to have the plant started and stopped automatically according to some predetermined schedule.

Time switches are available driven from the a.c. mains by a synchronous motor through a gearbox giving, usually, 24-hour rotation of a shaft. The shaft may hold two or four tappets which alternately make and break switch contacts. This gives one or two "on" periods per day. The tappets are adjustable either continuously or in $\frac{1}{4}$-hour steps. These time switches are usually intended for use with domestic central heating systems. A manual switch or switches are provided which permit the choice of a number of "programmes", so that either space heating or domestic water heating or both may be off or on, either continuously or under time control. The shaft usually carries a graduated scale, marked 1-24, to act as a 24-hour clock, giving a visible indication of the current time and of the settings of the tappets, which are carried round with the scale.

An alternative arrangement is one incorporating an electronic clock with digital display. The 24-hour day is divided into twelve time segments ranging from 1 to 6 hours long. Twelve three-position switches allow the choice of hot water only, heating and hot water on or off for each time segment. In addition, both heating and hot water can be chosen to be off continuously, on continuously or under time control by two overriding manual switches.

More sophisticated versions of the basic time switch are available, mainly for use in commercial and industrial premises. These have the facility of separate time settings for each day of the week. This permits a shutdown at weekends or early on a Saturday, and an early switch-on on Monday morning to bring the building up to the required temperature at a suitable time. A disadvantage of such a system is that the switch-on times are predetermined and are adhered to by the clock, whatever the weather conditions. The switch-on time must be chosen with regard to the worst expected weather to ensure that the building will be warm enough when required. This means that the building will usually be at the desired temperature earlier than necessary and before the working occupants arrive. This is wasteful of fuel. Similarly it may be possible, if the weather is unusually warm, to switch off the heating system earlier than usual. This facility is not available with an ordinary time switch. The next section describes equipment which will allow the morning switch-on to be delayed and the afternoon switch-off to be brought forward when weather conditions permit; the so-called optimum-start controller or optimiser.

3.6.1 Optimum Start Controllers

Figure 3.25(a) shows the temperature and heat-output profiles which would occur for a building and heating plant which possessed zero thermal capacity or heat storage, required to be kept at a given inside design temperature for a set period of occupancy. The building response to a heat input would be immediate. Real buildings do possess thermal capacity in their fabric and heating plant also takes time to come to full output after switch-on. Figure 3.25(b) shows that in order to bring a building up to the required temperature at the start of the occupied period, it is necessary to switch on the plant some time in advance. Obviously the higher the outside temperature the longer the switch-on can be delayed, as indicated in Fig. 3.25(c), thus saving fuel. There are three main factors which determine the switch-on time: the outside temperature, the inside temperature and the heating-up profile of the building/plant combination. In the earlier types of optimum start programmers the characteristics of the building and plant were embodied in the shape of a cam which was carried on the shaft of an ordinary 24-hour time switch. Different shapes of cam profile could cater for the responses of heavy or lightweight buildings, but once the cam was chosen this characteristic was fixed. In the recent versions of these units, known as optimisers, the outputs of inside and outside temperature sensors are fed in continuously to a micro-processor, which can take the decision to switch on the heating taking into account the previous history of the building response. The optimiser is adaptive that is, it is able to update its model of the building response in the light of past success or failure. The unit can be programmed for up to 12 months ahead to

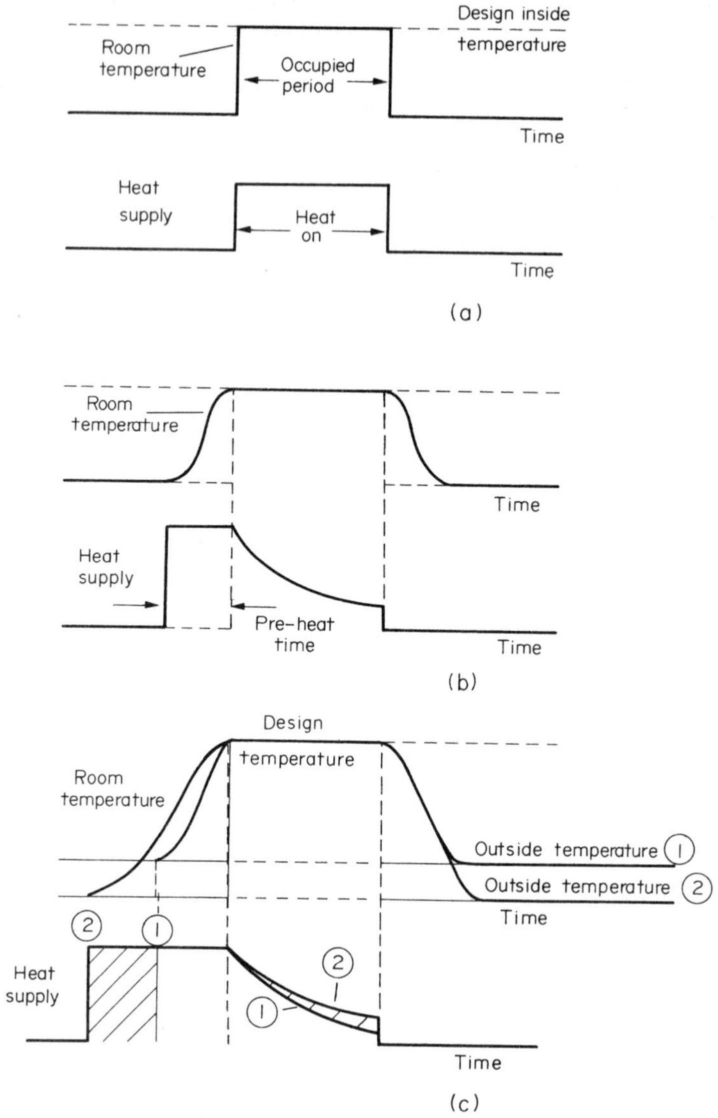

Fig. 3.25. Temperature and heat supply profiles. (a) Ideal building and plant with zero thermal capacity. (b) Real building and plant requires pre-heating to give required temperature at the start of occupied period. (c) With two constant levels of outside temperature, the switch-on time is modified. The shaded areas represent energy savings due to higher outside temperature.

include holidays, weekends and planned shutdown periods. A number of different occupancy patterns can also be included. When the pre-heat period is over, the building is maintained at the set temperature on normal thermostatic control.

If a certain permitted temperature drop at the beginning and end of the occupied period can be allowed to occur, the plant can be switched on later and switched off earlier, producing further savings as indicated in Fig. 3.26.

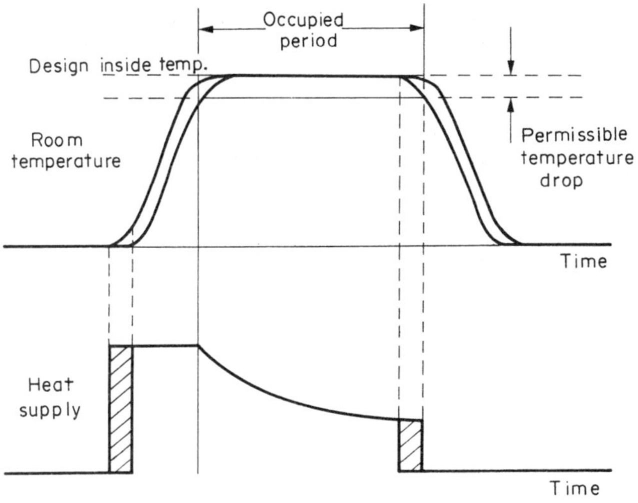

Fig. 3.26. If a permissible temperature drop in the occupied period is specified, the heating can be switched on later and off earlier. The shaded areas represent energy savings due to the permissible temperature drop.

During the unoccupied periods, the inside temperature is continuously monitored and if it falls below a given level the plant may be restarted to give protection against frost or condensation. If the mains supply to the unit fails or is temporarily disconnected, an internal battery maintains operation for 2 to 3 days.

3.7 THERMOSTATIC RADIATOR VALVES

These valves are applied in medium and low-temperature hot-water heating systems, to provide individual room-temperature control. The valve carries within it a complete self-operated control system: sensor, comparator, controller and actuator. The sensor consists of a capsule which is either filled with a wax which varies in volume with temperature change, or is a vapour-pressure system, being partially filled with a volatile substance whose saturated vapour-pressure changes with temperature. Although the sensor is mounted within the body of the radiator valve, it is thermally isolated from the valve body and from the pipework, so that its temperature is mainly dependent on the ambient room temperature. When a rise in ambient temperature is sensed, the capsule or bellows expands against a setting spring, the compression of which is chosen by the position of a knob, and moves the valve plug towards its seat, throttling the water flow. The valve must be mounted so that room air can pass over the sensor without obstruction. If this is not possible, perhaps because of curtains obstructing the air stream, a valve with a separate sensor may be used, connected by a capillary tube. Valves are also available with remotely mounted adjustment so that the room temperature may be adjusted at a convenient position.

Because the stroke of the valve is a continuous function of the temperature deviation, the control action is effectively the same as proportional control. The proportional band is usually about 2°C.

CHAPTER 4

CONTROLLER CHARACTERISTICS

Referring to Fig. 5.2, which shows the block diagram of a simple control system, the block labelled CONTROLLER provides a transfer function whose input signal is the error, the difference between the reference signal and the feedback signal. The controller must operate upon this input in some way so as to generate a suitable output signal to feed to the valve or other final control element. In this chapter we shall examine some of the types of transfer function which are used and the principles of operation of the mechanisms which embody them.

4.1 PROPORTIONAL CONTROL

In the case where the output signal of the controller is a continuous function of the input, the simplest relationship between them is a gain constant K, whereby the changes in input are multiplied by a suitable constant value to produce the output change. The gain constant cannot normally be a pure number since the nature of the output signal is usually different from that of the input. It is only for pneumatic controllers that the two signals are of the same type.

Generally the controller gain K is an adjustable parameter whose value can be chosen and set by the operator. Figure 4.1 shows the relationships between input and output for various settings of K. In theory the output signal may apparently take on any value but in practice its range is limited by the finite range of the final control element: a valve or damper can go only from fully closed to fully open. This fact can be regarded as a saturation effect. It produces the modified input/output curves as shown in Fig. 4.2, where the slope of the curves has been reversed to give the appropriate action for a heating duty. It will be convenient to discuss the operation of controllers in these terms, although in other cases, such as a cooling duty, it may be required that the valve opens on an increase in the controlled variable. This also depends on the selection of valve type.

In general, a proportional controller has to have two variables which can be set by the operator by means of front panel switches or knobs: the set point, that is the value of the controlled variable which the controller is set to maintain, and proportional gain, that is in effect the slope of the input/output curve. Practical controllers can easily be provided with a temperature scale and setting knob for the set point, but the proportional gain is normally replaced by an alternative, known as the Proportional Band. Other terms are "throttling range" and "modulating range". The proportional band is illustrated in Fig. 4.3, and may

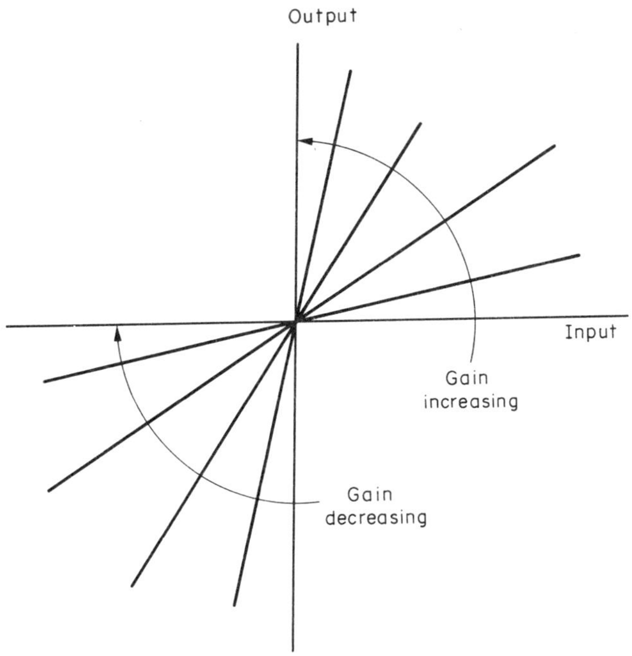

Fig. 4.1. Input/output characteristics for a proportional controller with various values of proportional gain constant.

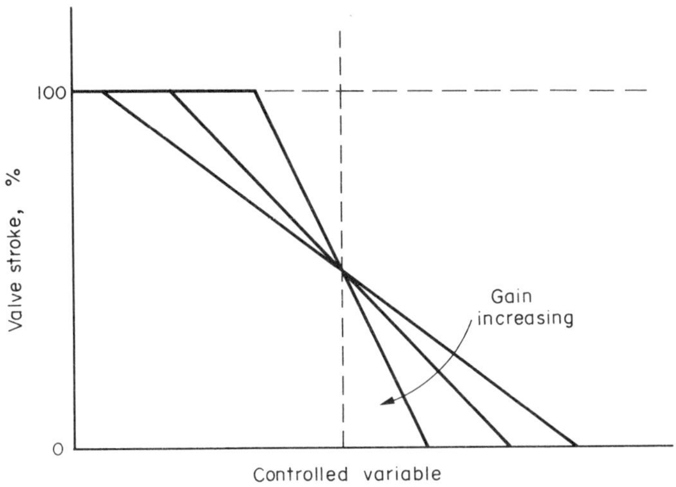

Fig. 4.2. Proportional controller characteristics with saturation by limited travel of valve.

be defined in two alternative ways.

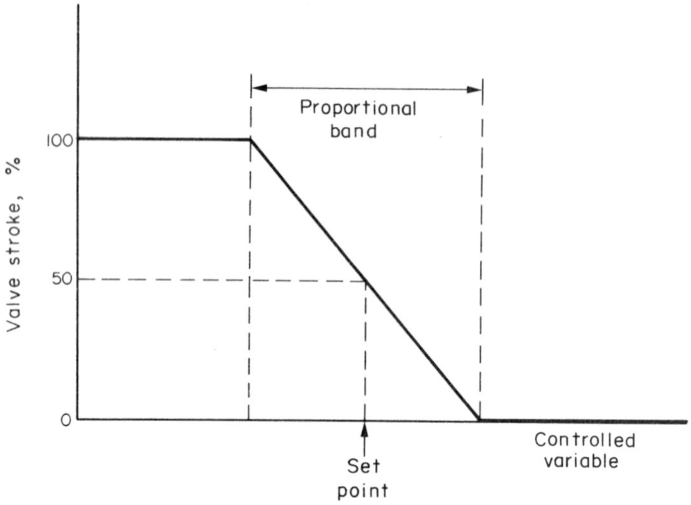

Fig. 4.3. Proportional band.

(i) As implied in Fig. 4.3, the proportional band may be defined as the change in the controlled variable required to make the final control element move through its full operating range or stroke.

(ii) Controllers are often scaled in terms of percentage proportional band, with a range of settings from about 2% up to perhaps 500%. The definition is then related to the span of the controller, i.e. the difference between the maximum and minimum set point settings.

$$\text{Percentage Proportional Band} = \frac{\text{proportional band as defined above}}{\text{span of controller}} \times 100\%.$$

Note that the numerator and denominator of the above fraction must be in the same units. For pneumatic controllers, where the same controller may be used to control any one of a number of different variables, proportional band may be defined as:

$$\text{Percentage Proportional Band} = (\Delta P_2/\Delta P_1) \times 100\%.$$

where ΔP_2 = change in output air pressure,

and ΔP_1 = change in input air pressure.

without saturation of the controller.

EXAMPLE 4.1. A proportional temperature controller has a range of set-point settings from 40 to 120°C. It operates an electric heater battery whose maximum output goes from 5.4 to 2.4 kW as the controlled temperature rises from 60 to 85°C. Find the proportional band and the percentage proportional band.

The controller span is 120-40 = 80°C.

The slope of the controller characteristic is

$$\frac{5.4-2.4}{85-60} = 120\ \mathrm{W/^oC}.$$

Since the maximum heater output is 15 kW, the proportional band is

$$15{,}000/120 = 125^oC.$$

The percentage proportional band is given by

$$(125/80) \times 100 = 156\%.$$

One important characteristic of proportional control is the existence of offset, as described briefly in Section 3.4. Offset is a steady-state difference between the set point and the actual value of the controlled variable. The offset depends on the load on the system and is an unavoidable consequence of proportional control action. It can be reduced by making the proportional band narrower although, as we shall see in Chapter 5, this can produce unstable operation if the proportional band is reduced too much.

The set-point scale on proportional controllers is often calibrated so that the offset is zero when the output of the final control element is at 50% of its maximum, as shown in Fig. 4.3. This is the normal assumption, although some manufacturers arrange that the offset is zero when the output of the final control element is at its maximum value.

One convenient illustration of a proportional controller with offset is the simple ball-valve with a float, installed in a cold-water cistern. If we regard the valve as a level controller, the set-point setting is chosen by the height at which the valve is installed in the wall of the tank. When no water is drawn from the tank, the level is constant and the float keeps the valve shut off. If water is drawn from the tank at a constant rate, the level falls so that the valve opens and water is allowed to flow into the tank to replace the draw-off. The fall in level may be regarded as the offset, and the water inflow is proportional to the value of the offset, as in Fig. 4.4. In this case the set point may be regarded as being at the zero valve stroke point.

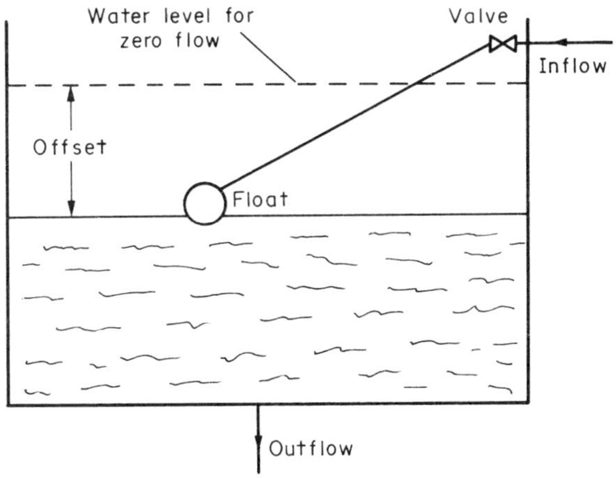

Fig. 4.4. A ball valve as a proportional level controller.

4.2 STEADY-STATE RESPONSES IN PROPORTIONAL CONTROL

In terms of the temperature control of a heated room, it is possible to regard the room heat-loss characteristic as being complementary to the proportional control characteristic. From this it is possible to predict the steady temperature which will exist for a given load, depending on the set point and proportional band. For these purposes the valve or final control element may be regarded as being included in the controller. The simplifying assumption of an ideal sensor will also be made. The resulting block diagram is shown in Fig. 4.5.

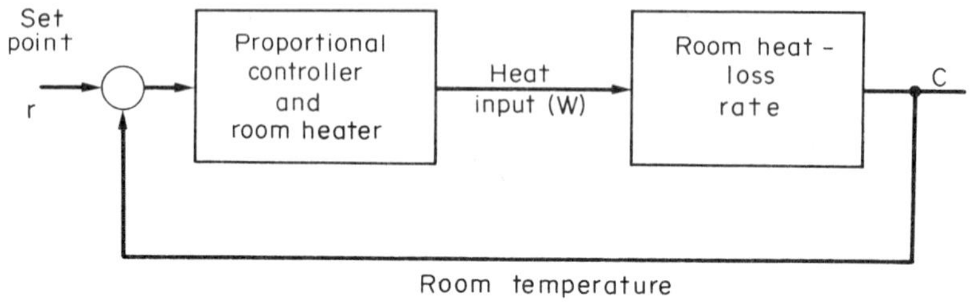

Fig. 4.5. Steady-state proportional control of room temperature.

The characteristics of the two blocks, connected by the heat input to the room, are as shown in Fig. 4.6. The simultaneous equations of the two characteristics are easily obtained by substitution in the general equation of a straight line.

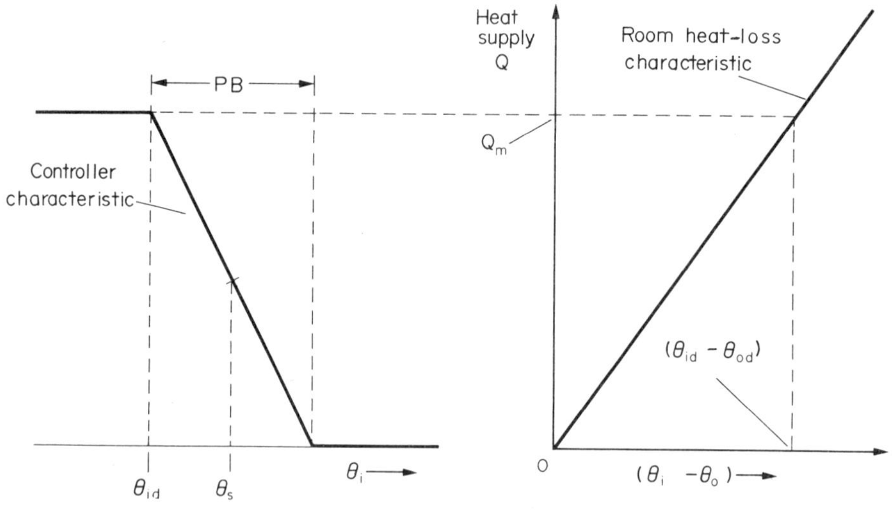

Fig. 4.6. Characteristic lines for controller and room heat loss.

The controller characteristic for heat inputs between the minimum, zero, and the maximum, Q_m, can be written as

$$Q = Q_m\,(\theta_s/PB + \tfrac{1}{2}) - Q_m\,\theta_i/PB \tag{4.1}$$

where Q = steady rate of heat supply to the room, W,

Q_m = maximum rate of heat supply to the room, W,

θ_s = set point, ^oC
PB = proportional band, ^oC,
θ_i = internal room temperature, ^oC.

This can be rearranged in terms of the ratio of actual heat-supply rate to maximum heat-supply rate

$$Q/Q_m = (\theta_s - \theta_i)/PB + \tfrac{1}{2}. \tag{4.2}$$

Note that equations (4.1) and (4.2) are valid only for room temperature in the range $(\theta_s\text{-}PB/2) < \theta_i < (\theta_s + PB/2)$.

The steady-state heat-loss characteristic of the room is determined by its insulation, ventilation rate, area, etc. The maximum rate of heat supply Q_m produces a steady-state temperature rise $\theta_{id} - \theta_{od}$ where θ_{id} is the design value of inside temperature and θ_{od} is the design value of outside temperature. The characteristic line passes through zero so its equation can easily be written as

$$Q = \frac{Q_m\,(\theta_i - \theta_o)}{(\theta_{id} - \theta_{od})} \tag{4.3}$$

$$= Q_m\,\Delta\theta/\Delta\theta d$$

where $\Delta\theta = \theta_i - \theta_o$, the actual value of inside-outside temperature diffference, ^oC,

and $\Delta\theta_d = \theta_{id} - \theta_{od}$, the design value of inside-outside temperature difference, ^oC.

Similarly, rearranging in terms of the ratio gives

$$Q/Q_m = \Delta\theta/\Delta\theta_d. \tag{4.4}$$

Substituting equation (4.4) in (4.2) eliminates the ratio Q/Q_m:

$$(\theta_s - \theta_i)/PB + \tfrac{1}{2} = (\theta_i - \theta_o)/\Delta\theta_d$$

and solving for the internal room temperature θ_i gives

$$\theta_i = \frac{2PB\theta_o + 2\theta_s\,\Delta\theta_d + PB\Delta\theta_d}{2\Delta\theta_d + 2PB}\,. \tag{4.5}$$

Rearranging the same equation for the outside temperature θ_o we have

$$\theta_o = \frac{2(\Delta\theta_d + PB)\theta_i - (2\theta_s + PB)\Delta\theta_d}{2PB} \tag{4.6}$$

We can now find the range of outside temperatures for which control will be maintained over the inside temperature. The maximum value of θ_i within the control band is $\hat{\theta}_i = \theta_s + PB/2$. This is the point at which the heating-control valve is just fully closed, so the heat supply will then be zero and the inside temperature will be equal to the outside temperature. The corresponding value of θ_o, $\hat{\theta}_o$, will thus also be equal to $\theta_s + PB/2$, which can be confirmed by substitution of $\theta_i = \theta_s + PB/2$ in equation (4.6).

The minimum value of θ_i within the control band is $\check{\theta}_i = \theta_s - PB/2$. At this point the heating valve will now be just fully open. The heat supply will be at its maximum and will maintain the design temperature difference between inside and outside, $\Delta\theta_d$. The corresponding outside temperature at which this will occur must therefore be

$$\check{\theta}_o = \theta_i - \Delta\theta_d$$
$$= \theta_s - PB/2 - \Delta\theta_d.$$

This again can be verified by substitution of $\theta_i = \theta_s - PB/2$ in equation (4.6). The range of outside temperature for which control will be maintained over θ_i is then

$$\Delta\theta_o = \hat{\theta}_o - \check{\theta}_o$$
$$= \theta_s + PB/2 - (\theta_s - PB/2 - \Delta\theta_d)$$
$$= \Delta\theta_d + PB \qquad (4.7)$$

which is an interesting and perhaps an unexpected result. Typically the design values of inside and outside temperatures in Britain are about 21 and -1°C respectively. With a proportional band of, say, 4°C, the range of outside temperatures for which control over θ_i will be maintained is then 26°C. For a set point of 19°C the values of $\hat{\theta}_o$ and $\check{\theta}_i$ are then 21 and -5°C respectively. These relationships are shown schematically in Fig. 4.7

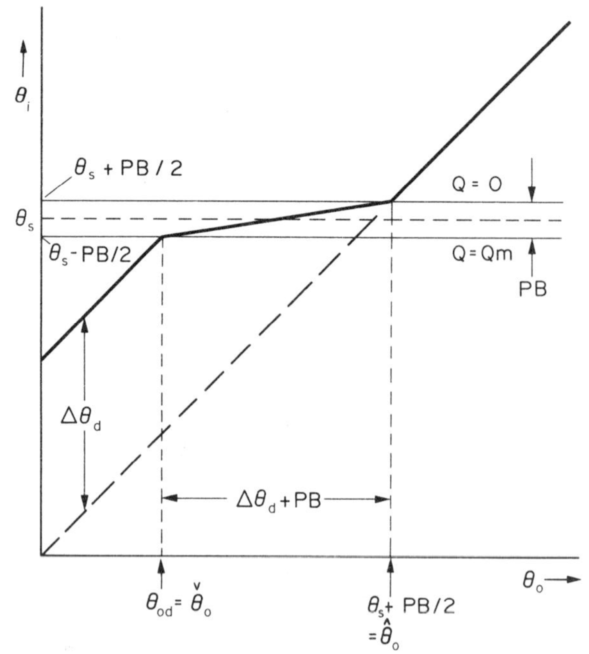

Fig. 4.7. Inside and outside temperature relationship with proportional control.

By substituting equation (4.5) into equation (4.1), the heat-supply rate for any value of θ_o in the range can be obtained:

$$Q = \frac{Q_m (2\theta_s + PB - 2\theta_o)}{2(\Delta\theta_d + PB)} \tag{4.8}$$

for $(\theta_s - PB/2 - \Delta\theta_d) < \theta_o < (\theta_s + PB/2)$.

It should be emphasised that it is assumed throughout this section that steady temperatures exist and the responses are all stable.

EXAMPLE 4.2. A building has a steady-state heat requirement of 400 W per degree temperature difference between inside and outside. The inside temperature is controlled by a proportional controller whose set point is 20°C and proportional band 2°C. The design temperature difference is 20°C. Find the inside temperature when the outside is steady at (a) + 10°C, (b) + 25°C, and (c) - 4°C. Also find the heat-supply rate when the outside temperature is + 5°C.

It is possible to solve this problem algebraically using equations (4.5) and (4.6) directly, but it is probably simpler and less prone to error if a graphical solution is used.

The relevant factors are as follows

$$\theta_s = 20, \ PB = 2, \ \Delta\theta_d = 20$$

and from these we can obtain the four coordinates of the two points at the extremes of the proportional band.

$$\hat{\theta}_i = \theta_s + PB/2 = 21^\circ C,$$

$$\check{\theta}_i = \theta_s - PB/2 = 19^\circ C,$$

$$\hat{\theta}_o = \hat{\theta}_i \text{ (when } Q = Q_m)$$

$$= 21^\circ C,$$

$$\check{\theta}_o = \theta_s - PB/2 - \Delta\theta_d$$

$$= 20 - 1 - 20$$

$$= -1^\circ C.$$

These points are plotted to scale in Fig. 4.8 (points B and C). A straight line joins B and C to give the control within the proportional band. For outside temperatures higher than 21°C the heating valve will remain fully closed and the slope will be unity, giving the line CD. Similarly, for outside temperatures lower than -1°C, the valve will remain fully open and the heat supply will be constant. The line AB will therefore also have a unity slope, with θ_i falling by one degree for each degree fall in θ_o. The contour ABCD thus gives the relationship between θ_i and θ_o both inside and outside the control band.

From Fig. 4.8, the inside temperatures when θ_o is +10, +25 and -4°C are 20, +25 and +16°C respectively. Note that the second and third cases lie outside the control band, which might be overlooked in a purely algebraic solution.

When the outside temperature is +5°C, the inside is seen to be 19.5°C, giving a temperature difference of 14.5°C and a steady-heat supply rate of

$$400 \times 14.5 = 5800 \text{ W}.$$

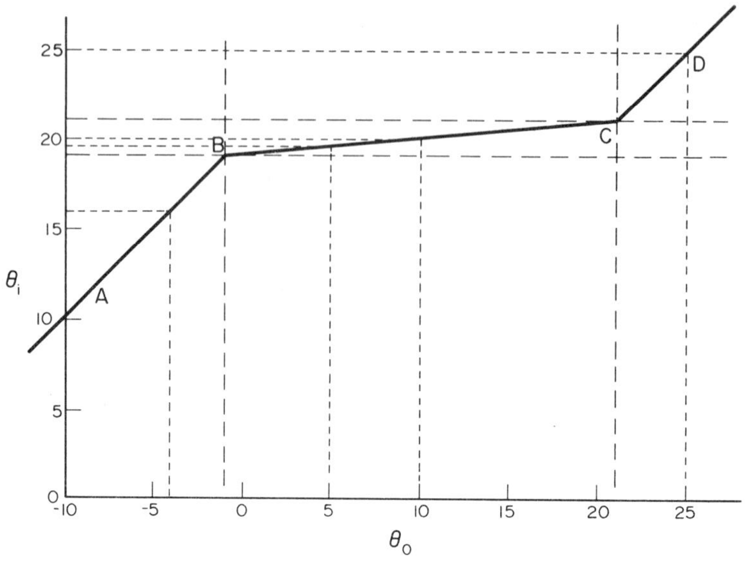

Fig. 4.8. Graphical solution to Example 4.2.

4.3 THREE-TERM CONTROLLERS; PROPORTIONAL, INTEGRAL AND DERIVATIVE ACTIONS

It appears that, when any given value of set point is set on a proportional controller, the controlled variable has that same value only for one value of the load. In all other cases there is a non-zero offset and a difference exists between the set point and the controlled variable. In many cases the proportional band is relatively narrow and the offset is acceptably small. In cases where the load is steady for long periods it may be possible to adjust the set point so that the controlled value becomes equal to the desired value, but this is not often convenient. If close control is necessary, the proportional band can be reduced, but this is equivalent to increasing the gain and as is shown in Chapter 5 this can produce oscillation and instability. These effects set limits on the accuracy of control available with proportional action only, and if better performance is required it is necessary to resort to some alternative or additional form of control action.

The control functions which are available consist of additions to the basic proportional control to remove the offset and to speed up the response. They are called Integral action and Derivative action. Figure 4.9 illustrates the typical responses of processes with such controllers to a step change in load. In Figure 4.9(a) the load change occurring at time zero is shown. Figure 4.9(b) shows the typical response of a system with only proportional control: after a delay the controlled variable c starts to change, and the control action is able to bring it only partially back to the original value, some offset effect remaining in the steady state. In Figure 4.9(c) the effect of adding integral action is indicated. The integral term generates a control signal which is proportional to the time integral of the error signal, so that it depends on the area under the error curve. A continuous offset will therefore eventually produce a corrective action, which will continue to accumulate until it has reached zero. In order to speed up the response of the system to changes in load, a derivative action can also be added to

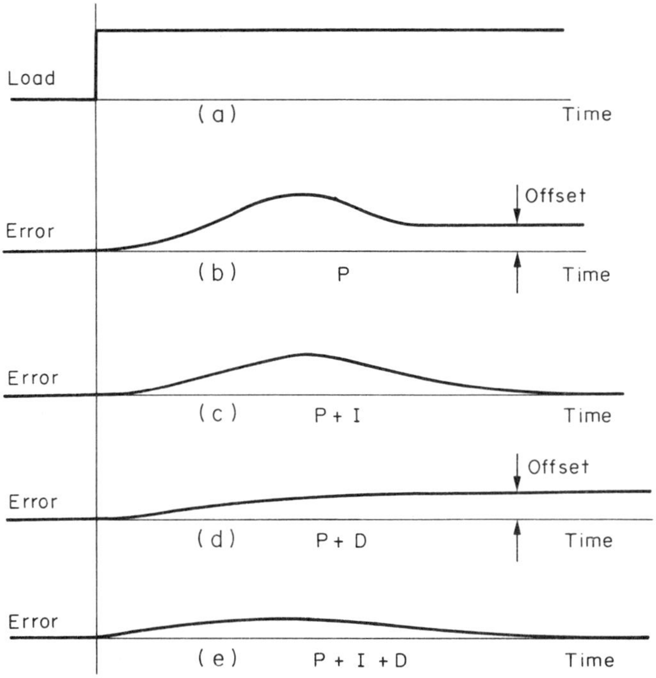

Fig. 4.9. Effects of Integral and Derivative actions. (a) Step change in load at time zero; (b) proportional control with offset. (c) proportional plus integral control; (d) proportional plus derivative control with offset; (e) proportional plus integral plus derivative control.

the proportional controller. This generates a control signal which is proportional to the time derivative or instantaneous rate of change of the error. This has no effect on the offset, where the slope is near zero, but it produces a larger initial response and earlier corrective action, as seen in Fig. 4.9(d). Some controllers combine all three types of action (proportional, integral and derivative: P+I+D) to include both advantages, as indicated in Fig. 4.9(e).

The relative strengths of the control terms have to be adjusted to match the characteristics and response of the process which is being controlled. The proportional gain constant K_p determines the sensitivity of the proportional action, but since the integral and derivative actions are time-related, they are adjusted by reference to so-called action times: the Integral Action Time T_i and Derivative Action Time T_d. These are defined in a manner illustrated in Fig. 4.10, where the controller input and output signals are assumed to be isolated from the control loop, so that the output cannot be fed back to the input and affect it. Figure 4.10(a) shows a step input of magnitude θ_1 applied to the input of a proportional controller. Ideally this produces a step function output signal of magnitude θ_2. The proportional gain K_p is then defined as the ratio θ_2/θ_1. The output signal is shown inverted to emphasise the negative feedback effect. In Fig. 4.10(b) a negative-going step of magnitude θ_1 is shown applied to the input of a P+I controller. The output signal consists of two parts: a positive-going step of magnitude $K_p\theta_1$, due to the proportional term, and a ramp of slope $K_p\theta_1/T_i$, due to the integral term.

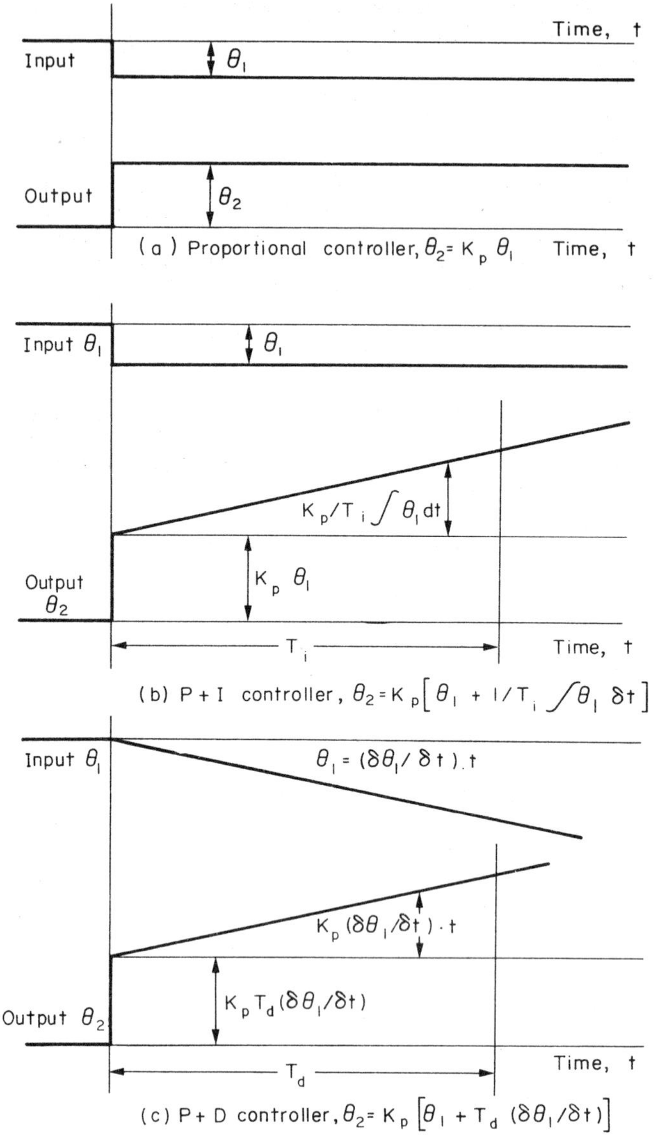

Fig. 4.10. Definitions of Integral action time and Derivative action time, adapted from BS 1523: Part 1: 1967.

$$\text{Input magnitude} = \theta_1,$$

$$\text{Output magnitude} = K_p\theta_1 + (K_p/T_i)\int \theta_1 dt,$$

$$= K_p\theta_1 + K_p\theta_1 t/T_i$$

since θ_1 is constant.

The portion of the output which is due to the integral term will increase uniformly with time, whereas the portion due to the proportional term remains constant. The integral action time T_i is defined as the time taken for the two terms to become equal.

$$K_p\theta_1 = K_p\theta_1 t/T_i.$$

This equation is satisfied when $t = T_i$.

In the case of the P+D controller, a step function applied to the input will in theory give an infinite response at the output. A ramp input of constant slope $\partial\theta_1/\partial t$ is employed instead, as shown in Fig. 4.10(c). The output signal again consists of two parts. The part due to the proportional term is a ramp and that due to the derivative term is a constant step since the input signal is of constant slope

$$\text{Input magnitude} = (\partial\theta_1/\partial t)\cdot t,$$

$$\text{Output magnitude} = K_p\,T_d\,(\partial\theta_1/\partial t) + K_p\,(\partial\theta_1/\partial t)\cdot t.$$

The derivative action time is defined as the time taken for the two parts of the output signal to become equal.

$$K_pT_d\,(\partial\theta_1/\partial t) = K_p\,(\partial\theta_1/\partial t)t.$$

This equation is satisfied when $t = T_d$.

To summarise, the input/output equations and transfer functions for the various forms of controller available are as follows. The input variable is θ_1, the output is θ_2.

Proportional (P)

$$\theta_2(t) = K_p\,\theta_1(t), \qquad \theta_2(s) = K_p\,\theta_1(s)$$

since K_p is a constant.

Proportional plus integral (P+I)

$$\theta_2(t) = K_p\,\theta_1(t) + (K_p/T_i)\int\theta_1(t)\,d_t,$$

$$\theta_2(s) = [K_p + K_p/T_i s]\cdot\,\theta_1(s).$$

Proportional plus derivative (P+D)

$$\theta_2(t) = K_p\theta_1(t) + k_pT_d\,(\partial\theta_1/\partial t),$$

$$\theta_2(s) = [K_p + K_p\,T_d s]\cdot\,\theta_1(s).$$

Proportional plus integral plus derivative (P+I+D)

$$\theta_2(t) = K_p\theta_1(t) + (K_p/T_i)\int\theta_1(t)dt + K_pT_d\,(\partial\theta_1/\partial t),$$

$$\theta_2(s) = [K_p + K_p/T_i s + K_pT_d s]\theta_1(s).$$

The range of adjustment of T_i is generally from about 1 minute to about 60 minutes. The range of adjustment of T_d is generally from about ¼ minute to about 15 minutes.

4.4 STABILITY LIMITS WITH CONTINUOUS CONTROLLERS

It is demonstrated in Chapter 5 that increasing the value of K_p in a proportionally controlled system can produce instability if the order of the transfer function of the process is three or higher. The addition of integral and/or derivative action to give two-term or three-term controllers also affects the stability of the system. To demonstrate the effects on response, a simple example will be analysed.

EXAMPLE 4.3. Consider the simple system shown in Fig. 4.11 where F_c represents the controller transfer function. Initially we take a simple proportional

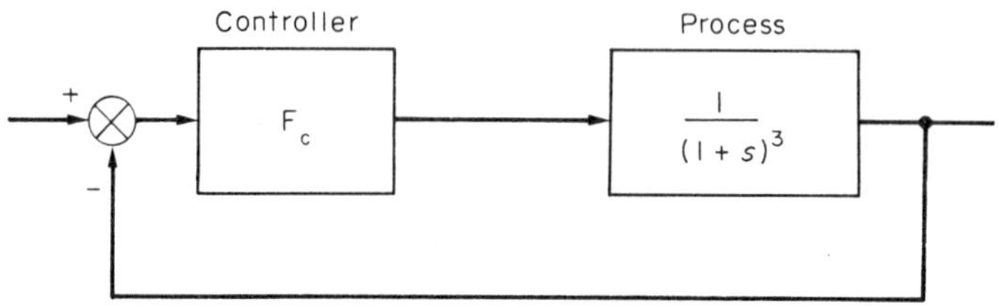

Fig. 4.11. General system for Example 4.3.

controller, so that $F_c = K_p$. To determine the limiting value of K_p for stable response we can employ the Routh-Hurwitz criterion, as described in Section 5.8. The characteristic equation is

$$1 + K_p/(1+s)^3 = 0$$

or

$$s^3 + 3s^2 + 3s + (K_p + 1) = 0$$

which gives the following array:

s^3	1	3
s^2	3	$K_p + 1$
s^1	$(8-K_p)/3$	0
s^0	$K_p + 1$	

from which it is seen that the s^1 term in the first column will become negative if K_p is greater than 8.

If we now choose to let $K_p = 5$, which is well within the bounds of stability for a proportionally controlled system, and add integral action, the characteristic equation now becomes

$$1 + 5\,(1 + 1/T_i s)/(1 + s)^3 = 0$$

or

$$T_i s^4 + 3T_i s^3 + 3T_i s^2 + 6T_i s + 5 = 0$$

Note that the equation is now fourth order instead of third order. The Routh array is as follows:

s^4	T_i	$3T_i$	5
s^3	$3T_i$	$6T_i$	
s^2	T_i	5	
s^1	X	0	
s^0	5		

where $X = (6T_i^2 - 15T_i)/T_i = 6T_i - 15$, which is negative if $T_i < 2.5$. Hence, for a small value of T_i the addition of the integral term has rendered the system unstable. Note that a short integral action time means that the strength of the integral term is large, because T_i appears in the denominator.

Returning now to the original case, if we choose a value for K_p of 10, which produces instability, and add derivative action to the controller, the characteristic equation becomes

$$1 + 10\ (1 + T_d s)/(1 + s)^3 = 0$$

or

$$s^3 + 3s^2 + (3 + 10T_d)s + 11 = 0$$

which results in the following array:

s^3	1	$3 + 10T_d$
s^2	3	11
s^1	Y	0
s^0	11	

where $Y = (30T_d - 2)/3$, which is negative if $T_d < 1/15$. Hence the system will be stable if $T_d > 1/15$, in spite of the large value of proportional gain K_p.

This simple example illustrates the general effects of integral and derivative actions on stability. Integral action tends to destabilise the system; derivative action tends to reinforce the stability. Although integral action may be introduced in order to reduce or remove the offset of proportional control and give more accurate control, its effect may be to make the system oscillate and worsen the control. Although in theory the addition of derivative action to a proportional controller will not remove the offset, it will nevertheless allow a narrower proportional band to be chosen so that offset may be reduced to an acceptable value. Obviously it is necessary to choose suitable settings for K_p, T_i and T_d. The next section describes some simple empirical methods with which an approximate initial choice can be made.

The time response of many processes can be represented by that of a pure delay L with a single time constant T, and it is possible to determine the equivalent values of L and T from an experiment carried out on the process. Taking the assumption of such a process transfer function it is possible to predict the behaviour of the system when particular types of controller are applied. Figure 4.12 shows the block diagram which is to be assumed, where F_c represents the controller transfer function.

First considering proportional control only, then F_c will be the proportional gain K_p. It is convenient to distinguish between three types of response: stable

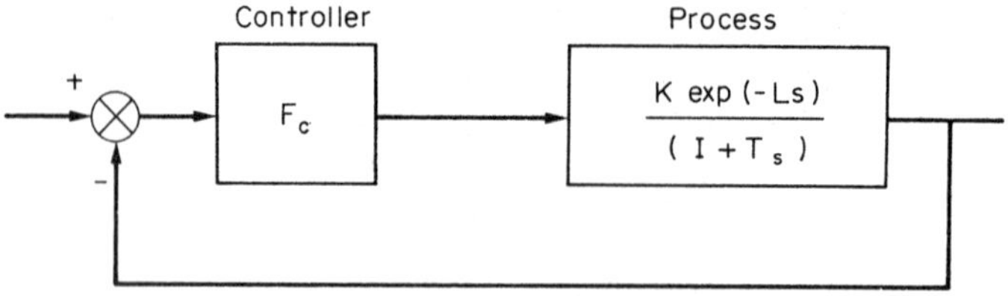

Fig. 4.12. Representation of a generalised process by a pure time delay and a single time constant.

and non-oscillatory, stable with a damped oscillation, and unstable. Figure 4.13 shows the boundaries between these response types, in terms of the overall gain K_pK and the ratio L/T.

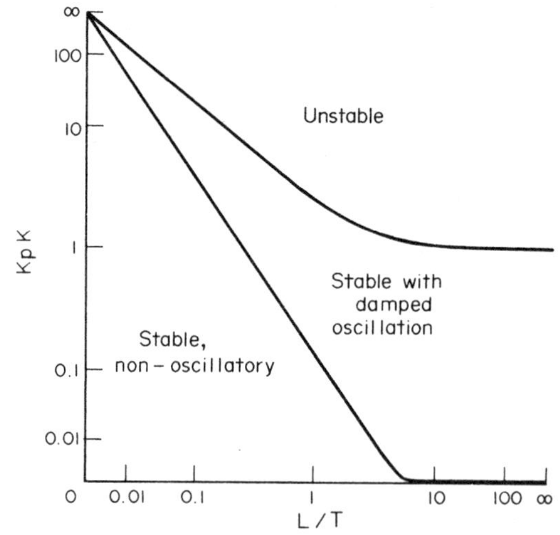

Fig. 4.13. Stability and oscillation limits for the system of Fig. 4.12 with proportional control (after Junker).

For proportional plus integral control an extra parameter, the integral action time T_i, is introduced, and it becomes unwieldy to show both damped and aperiodic responses. Figure 4.14 shows stability boundaries for various values of overall gain $K_p \cdot K$ in terms of the ratios T_i/L and L/T. The controller transfer function F_c is taken to be K_p $(1 + 1/T_is)$, appropriate for a P+I controller.

4.5 EMPIRICAL CHOICE OF CONTROLLER SETTINGS

In order to give some guidance on suitable initial settings for controllers, which will give reasonably good control without being either too sluggish or too oscillatory, a number of simple tests have been developed. These are more specifically aimed at the process industries such as petrochemicals, but there is no reason why they should not be used for other applications.

The assumption is that there is a closed-loop negative feedback control process

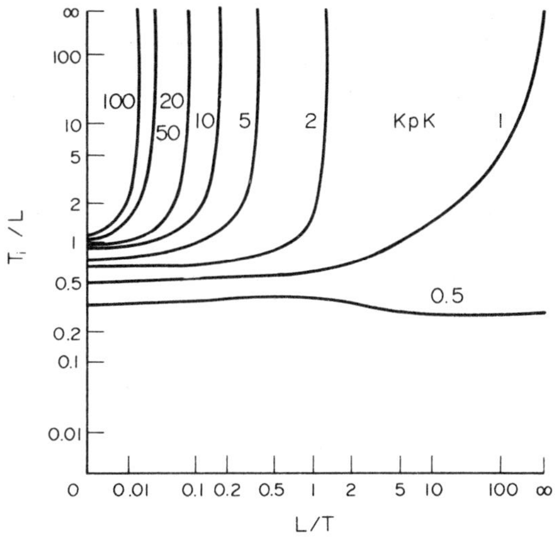

Fig. 4.14. Stability limits for the system of Fig. 4.12 with P+I control. The region below and to the right of each curve is the unstable region for the given value of K_pK (after Junker).

as in Fig. 4.15(a) where the characteristics of the valve, process and sensor are fixed and it is necessary to set the controller so as to match these characteristics satisfactorily.

1. *Process reaction curve method.* In this method the controller is isolated and the test is carried out on the remainder of the control loop. It must be arranged that a signal can be injected in place of the controller output and a signal observed or recorded which would enter the controller input. See Fig. 4.15(b). A sustained step input of magnitude x is applied to the valve, and the response at θ2 is observed and recorded, as in Fig. 4.16(a). In most cases of interest, the response will follow the general form of the S-shaped curve shown in 4.16(a), which is the resultant response of all the lags, delays, time constants and gain factors in the controlled system. From the experimental curve three values are measured, corresponding to a pure delay L, a single time constant T and a steady-state gain factor y/x. The equivalent overall transfer function is therefore

$$\frac{(y/x)\cdot \exp(-Ls)}{1 + Ts}$$

and it is possible to calculate the necessary controller settings to give a reasonably good control performance. The results published by two sets of workers are set out in Tables 4.1 and 4.2.

These responses are aimed at giving an oscillatory step response with a decay ratio of about $\frac{1}{4}$.

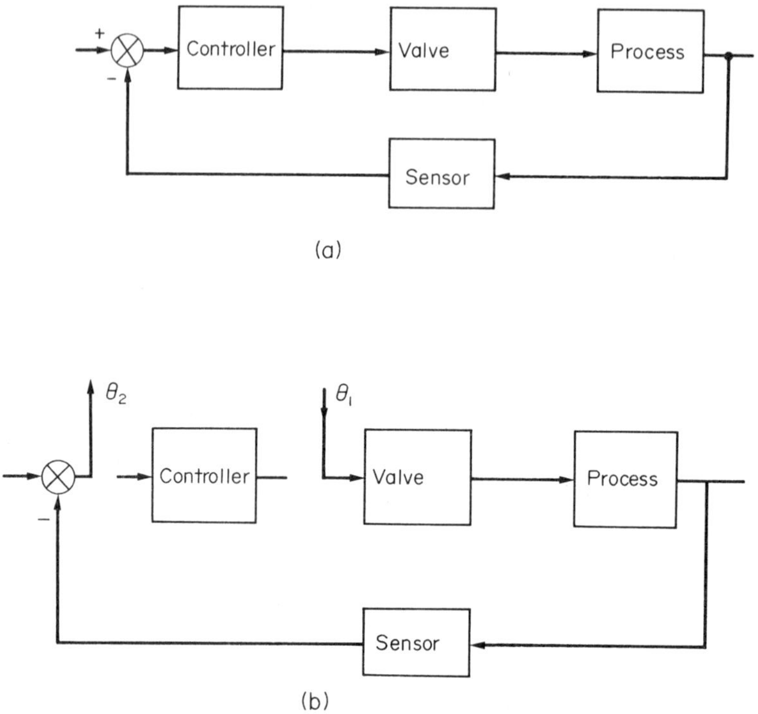

Fig. 4.15. System for empirical choice of controller settings: (a) basic feedback loop; (b) controller isolated for process reaction curve method.

Ziegler and Nichols settings

TABLE 4.1 (A = x/y, B = T/L, see Fig. 4.16(a)

Type of controller	K_p	T_i	T_d
P	A·B	-	-
P + I	0.9·A·B	3.3L	-
P + I + D	1.2·A·B	2L	0.5L

Cohen and Coon settings

TABLE 4.2 (A, B as before, R = L/T, see Fig. 4.16(a))

Type of controller	K_p	T_i	T_d
P	A·B(1+R/3)	-	-
P + I	A·B(1.1+R/12)	$L\,\dfrac{(30+3R)}{(9+20R)}$	-
P + D	A·B(1.25+R/6)	-	$L\,\dfrac{(6-2R)}{(22+3R)}$
P + I + D	A·B(1.33+R/4)	$L\,\dfrac{(32+6R)}{(13+8R)}$	$L\,\dfrac{4}{(11+2R)}$

2. *Ultimate cycling method*. An alternative empirical test which has been suggested is to allow the controller to remain connected in the control loop, but to decrease the proportional band progressively (corresponding to increasing the gain K_p) until the system just sustains steady oscillations of constant amplitude. The integral and derivative terms must first be disconnected from the controller or their effect should be reduced as much as possible by setting T_i to its maximum value and T_d to its minimum value. The smallest value of K_p which will just produce a steady oscillation without dying away is designated K_p'. The period of the oscillations is measured and is designated T'. The recommended controller settings are then related to K_p' and T' and are set out in Table 4.3.

Controller settings for the ultimate cycling method.

TABLE 4.3 (see Fig. 4.16(b))

Type of controller	K_p	T_i	T_d
P	$0.5\,K_p'$	-	-
P + I	$0.45K_p'$	0.8T'	-
P + I + D	$0.6\,K_p'$	0.5T'	0.125T'

4.6 ON/OFF CONTROLLERS

It will be shown in Chapter 6 how a non-linear system may be analysed in terms of the cycling time and amplitude. It will be convenient to deal here with some of the performance characteristics of simple systems controlled by on/off controllers. The phase plane will also be introduced here as a convenient way of illustrating the behaviour of such systems.

4.6.1 Thermostat with Zero Hysteresis Controlling a Single Time Constant Process

The controlled system is illustrated in Fig. 4.17, and may be thought of as a simple temperature-control system. K_1 is the rate of heat supply to the process, watts, and K_2 is the thermal resistance of the process, W/°C. From Fig. 4.17 we can write

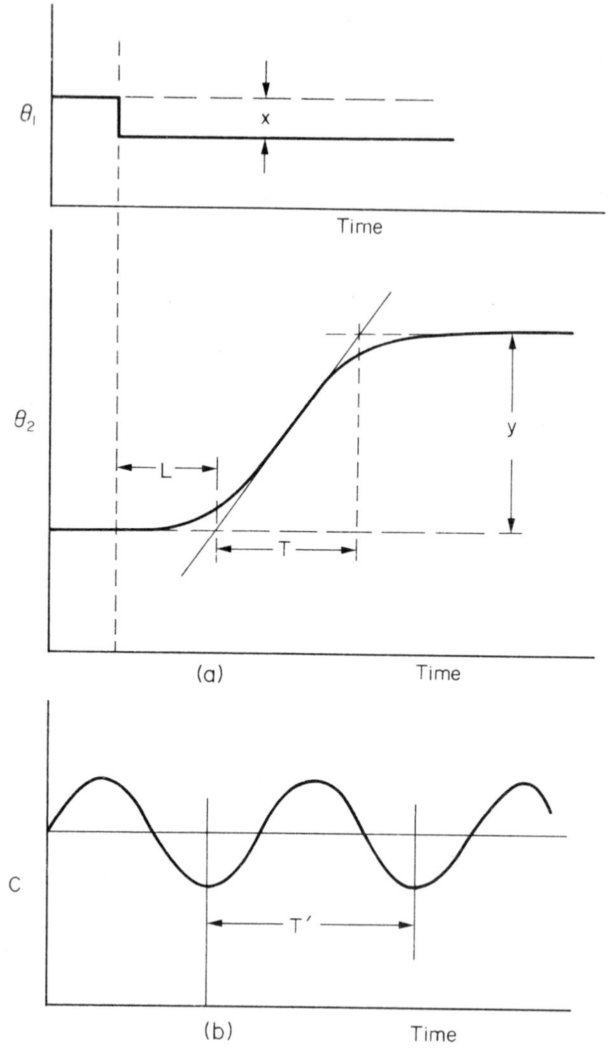

Fig. 4.16. (a) Input and output waveforms for process reaction curve method. (b) Continuous cycling method. (Adapted from Cohen and Coon.)

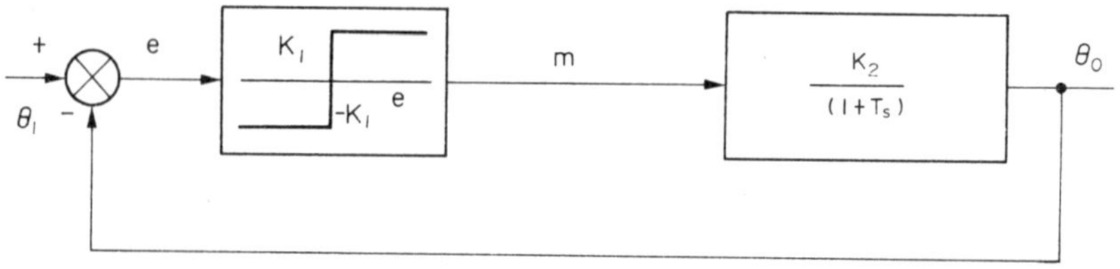

Fig. 4.17. On/off controlled system with zero hysteresis.

$$\theta_o(s) = K\ \mathrm{sgn}(e)/(1 + Ts) \tag{4.9}$$

where sgn(e) = + 1 for e > 0

and - 1 for e < 0,

$K = K_1\ K_2.$

If we choose to take $\theta_1 = 0$, we can write $e = -\theta_o$, so that

$$-e(s) = K\ \mathrm{sgn}(e)/(1 + Ts) \tag{4.10}$$

and taking s as a differential operator gives the differential equation

$$-e - T\ de/dt = K\ \mathrm{sgn}(e). \tag{4.11}$$

This equation cannot be solved directly because of the signum function, but it is possible to consider the two cases separately.

sgn(e) = +1

Equation (4.10) then becomes

$$-e - T\dot{e} = K \tag{4.12}$$

which is a straight-line equation in e and $\dot{e}$ with intercepts at -K and -K/T on the e and $\dot{e}$ axes respectively.

sgn(e) = -1

Equation (4.10) then becomes

$$e + T\dot{e} = K \tag{4.13}$$

which can be regarded as representing a similar straight line with intercepts at +K and +K/T.

These lines are shown in Fig. 4.18. The plot of e versus $\dot{e}$ is termed the phase plane plot, and time is a parameter which increases along the curve. The values of time can be reconstructed from the relationship

$$t_{AB} = \int_A^B (1/\dot{e})\cdot de \tag{4.14}$$

where t_{AB} is the time taken to traverse between two points A, B on the phase plane.

In Fig. 4.18, starting from a point such as A, the state point will travel to B as indicated by the arrows. At this point a switching will occur and the slope will instantly change from a positive to a negative value. The state point will instantly jump to C when another switching will occur. The state point will thus oscillate at (theoretically) infinite frequency between the two positions B and C.

4.6.2 Thermostat with Non-zero Hysteresis

If the control element in Fig. 4.17 is now replaced by a thermostat whose hysteresis value is not zero, such as that shown in Fig. 3.2(b), the switching points will occur at a distance h/2 away from the $\dot{e}$-axis. Equation (4.9) becomes

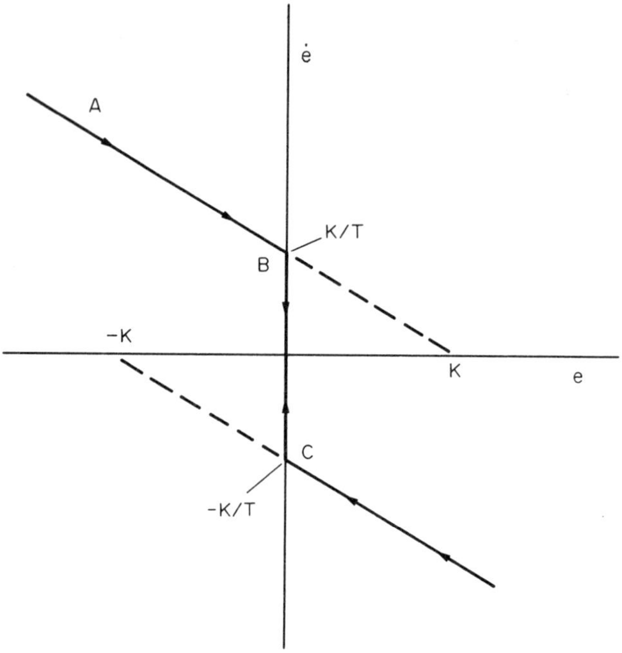

Fig. 4.18. Phase plane portrait of on/off system with zero hysteresis.

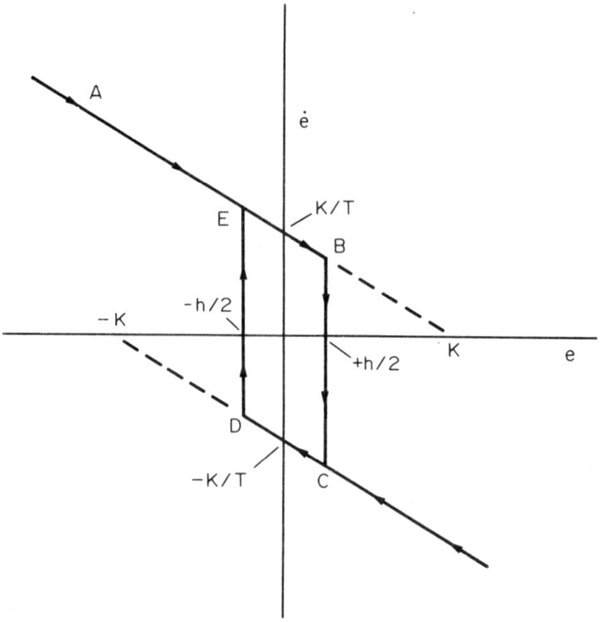

Fig. 4.19. Phase plane portrait of on/off system with total hysteresis h.

$$\theta_o(s) = K\,\mathrm{sgn}(e + h/2)/(1 + Ts) \tag{4.15}$$

and this leads to the phase plane picture shown in Fig. 4.19. Starting from cold at a point such as A, the state point must reach point B before a switching will occur to the point C. The point will then traverse CD when the next switching will occur. The closed path BCDE will be traversed continuously.

CHAPTER 5

BASIC THEORY OF LINEAR AUTOMATIC CONTROL

5.1 INTRODUCTION TO CLOSED-LOOP NEGATIVE FEEDBACK CONTROL

In order to introduce some of the basic ideas of automatic control theory we shall first consider a simple example of manual control. Figure 5.1 shows a hot-water calorifier whose secondary outlet temperature is being controlled by an operator. Before the operator can carry out his task satisfactorily, a number of requirements

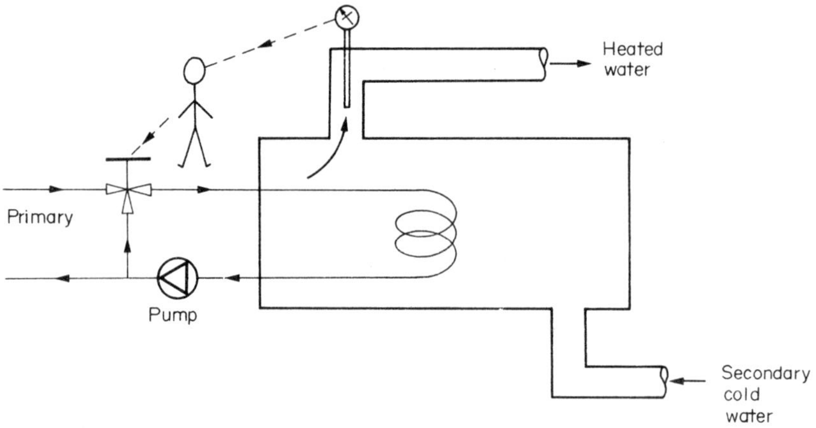

Fig. 5.1. A manually controlled system.

must be met:

1. He must be told what temperature is required for the water (set point or desired value).

2. He must be provided with some means of influencing the temperature (control element).

3. He must be provided with some means of observing the temperature (sensing element).

4. He must be instructed what to do to move the temperature in a desired direction (control function, negative feedback).

If the system is initially in a steady state with the required temperature achieved and some disturbance then occurs, this will cause the controlled temperature to deviate from its previous condition. The operator will not know anything of this until the pointer on the dial of the indicator moves and shows him that some change has occurred. This illustrates the principle that the precision of the control can only be as good as that of the sensor. When the pointer has moved the operator must then compare the current reading (the controlled variable c) with the required temperature (the set point r) and compute the difference between them, to obtain the error signal e:

$$e = r - c. \tag{5.1}$$

Note that the sign of e indicates whether the controlled temperature is too high or too low, which determines the direction of the corrective action required; whether to open up the valve or close it down. The size of e determines the amount of corrective action necessary. When the valve is turned in the correct direction by the correct amount, the temperature will eventually return to its original value.

It can be seen that there is a closed loop of signal or information which passes through the sequence: secondary water, temperature sensor, indicator, eye, brain, muscle, control valve, primary temperature and then secondary water again. The physical nature of the signal is different in various parts of the loop, but each part affects the next in turn and a change at any one point will propagate round the loop, as shown in Fig. 5.2, which shows the block diagram equivalent of Fig. 5.1.

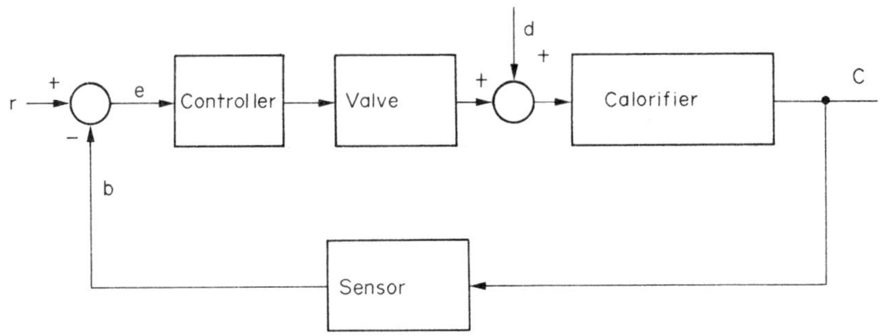

Fig. 5.2. Block diagram of Fig. 5.1.

There are three basic types of component used in block diagrams: the Block itself, the Summing junction and the Distribution point, as shown in Fig. 5.3. Each type of component has significance in mathematical terms. The block (Fig. 5.3(a)) has a single input and a single output. It contains a transfer function which operates on the input signal to generate the output. Input and output are distinguished by arrow heads. It will be shown later how the Laplace transform can be used to reduce the differential equation form of system dynamics to an algebraic form which is suitable for block diagram representation. The Summing junction (Fig. 5.3(b)) has two or more inputs and one output. The positive or negative signs placed next to the inputs indicate an algebraic relationship between the inputs to produce the output. The Distribution point (Fig. 5.3(c)) allows the same signal to be shared among a number of points.

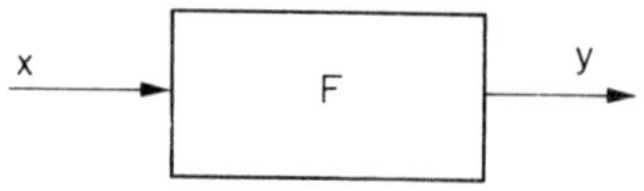

(a) Block: $y = x.F$

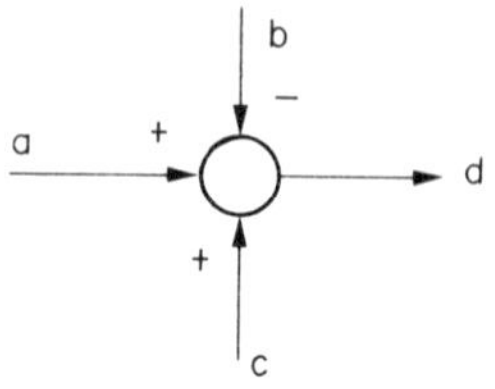

(b) Summing junction: $d = a - b + c$

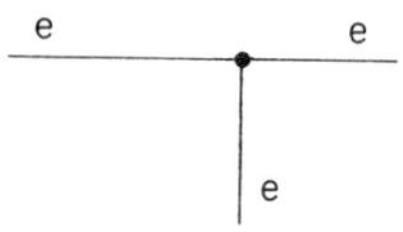

(c) Distribution point

Fig. 5.3. Block diagram components.

The components of the block diagram are joined by lines. The same signal exists at all points along each line: there is no loss in the way that voltage, for instance, will drop along an electrical cable.

Blocks may be joined in series or parallel or in series-parallel combinations. As shown in Fig. 5.4, a series arrangement implies a product of transfer functions, a parallel connection implies a sum of transfer functions.

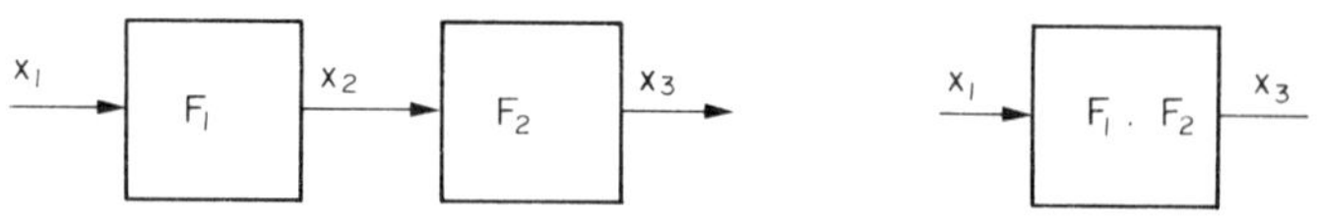

(a) Series: $x_3 = x_1\ (F_1 . F_2)$

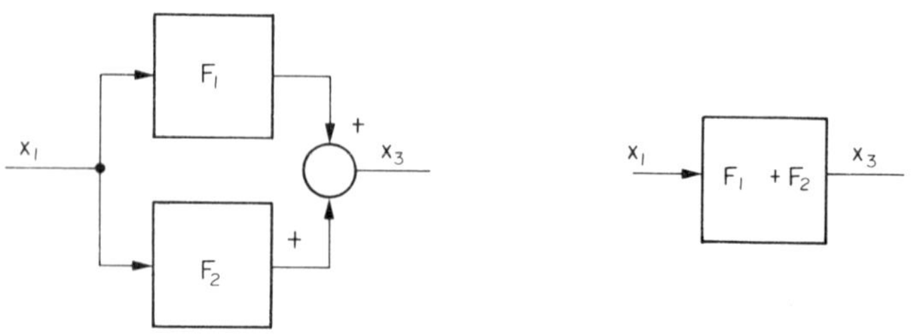

(b) Parallel: $x_3 = x_1\ (F_1 + F_2)$

Fig. 5.4. Series and parallel combinations of blocks.

Returning to Fig. 5.2, it can be seen that the controlled variable c is fed back by the sensor, and the summing junction on the left-hand side acts as a comparator. It compares the current value of c with the reference input r and forms the error signal e according to the equation

$$e = r - b \tag{5.2}$$

where b is the output of the sensor. Note that the negative sign at this summing junction is very important because it produces the *negative* feedback. The system is required to respond positively to change in r: an increase in the set point or reference input r should produce a corresponding increase in the controlled variable c. However, if the system is to be self-regulating, it must act so as to oppose undesired changes in c; such changes should be detected by the sensor which must cause a corrective action to be taken by the controller and valve. A positive sign associated with the feedback path would imply *positive* feedback whereby any change in the controlled variable would be reinforced, producing instability. The valve would be driven to an extreme position, fully open or fully closed, and would stay there. This effect is occasionally seen in practice when the motor connections are inadvertently reversed.

The reason why closed-loop control is desirable in this case is because of the presence of the disturbance input d shown in Fig. 5.2. This represents the heat content of the secondary cold water entering the calorifer. The temperature of the heated water is determined by this heat content, the flow rate and the heat output of the primary coil. Changes in the entry temperature of the secondary water will disturb the controlled variable and must be compensated for by modulating the heat output of the primary via the control valve.

5.2 TRANSFER FUNCTION DESCRIPTION OF LINEAR SYSTEMS

To make the block diagrams useful it is necessary to be able to write mathematical relationships between the variables. These are usually obtained by writing heat-balance or energy-balance differential equations for the components and applying the Laplace transform which converts the differential equations into an algebraic form so that a transfer function can be extracted.

To illustrate a simple application consider the system of Fig. 5.5, which shows a thermometer bulb immersed in a fluid. Using the D-operator we can derive a transfer function for this bulb.

Let the bulb contain V m^3 of mercury of specific thermal capacity (specific heat) C J/kg^{o}C and density ρ kg/m^3. The temperature of the surrounding fluid is θ_1 and the actual instantaneous temperature of the bulb is θ_2. The instantaneous reading of the thermometer is thus θ_2 and we wish to establish the functional relationship between θ_1 and θ_2.

Consider a heat balance on the bulb:

$$\text{Rate of heat entry} = \text{rate of heat storage (watts).} \tag{5.3}$$

In fact some heat will be conducted along the stem of the bulb but this is negligible. Writing expressions for the components of the heat balance:

$$(\theta_1 - \theta_2)h\ A = \rho CV\dot{\theta}_2 \tag{5.4}$$

where h = heat transfer coefficient at the surface, W/m^2 ^{o}C

A = exposed surface area of the bulb, m^2

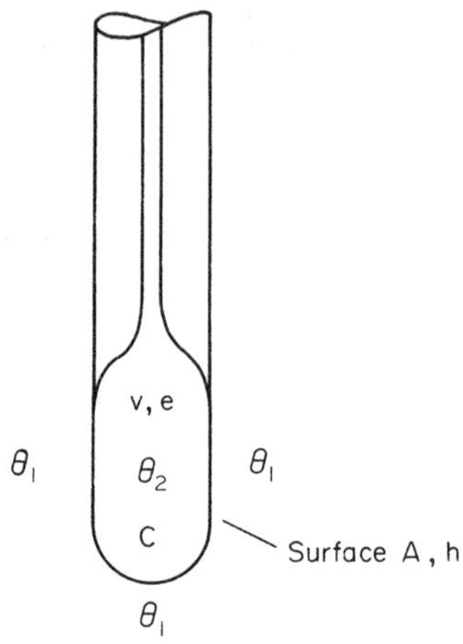

Fig. 5.5. Simple thermometer bulb.

and $\dot{\theta}_2$ indicates $d\theta_2/dt$, the first differential of θ_2 with respect to time.

Rearranging we have

$$\theta_1 = \theta_2 + \frac{\rho CV}{hA}\dot{\theta}_2. \tag{5.5}$$

Note that the units of the coefficient $\rho CV/hA$ are easily shown to be seconds: this is the time constant ofthe thermometer. Hence the equation can be written

$$\theta_1 = \theta_2 + T\dot{\theta}_2 \tag{5.6}$$

where $T = \rho CV/hA$ (5.7)

and applying the D-operator ($D = d/dt$):

$$\theta_1 = \theta_2 + TD\theta_2 = (1 + TD)\theta_2, \tag{5.8}$$

$$\theta_2 = \left[\frac{1}{1 + TD}\right]\theta_1. \tag{5.9}$$

The expression in brackets is the sensor transfer function. It enables us to predict the response θ_2 of the sensor output to an input function θ_1. For most purposes it is more convenient to use instead of D the Laplace operator s defined by the Laplace transform.

5.3 LAPLACE TRANSFORMS

The Laplace transform is one of the class of integral transforms which can be used to simplify the solution of linear differential equations by converting the differential equation to an algebraic equation. The solution procedure is systematic and initial conditions are easily included.

The transform is defined by the integral:

$$f(t) = \int_0^\infty e^{-st} f(t)\cdot dt = F(s). \tag{5.10}$$

The integral may be regarded as a rule for converting f(t), a function of time t, to a corresponding function $F(s)$ of the complex variable $s = \sigma + j\omega$.

In order to demonstrate the construction of a transfer function from a differential equation, we take a general linear second-order equation:

$$a\ddot{\theta}(t) + b\dot{\theta}(t) + c\theta(t) = f(t) \tag{5.11}$$

where a, b and c are real numerical constants, $\theta(t)$ is the time function of the response (the output variable) which we seek as a solution, and f(t) is the time function of the excitation or input variable which produces the response. The dot notation is used whereby $\dot{\theta}(t)$ denotes $d\theta/dt$ and $\ddot{\theta}(t)$ denotes $d^2\theta/dt^2$. Applying the transform to each term, and integrating by parts where necessary:

$$\text{(i)} \quad L\left[a\ddot{\theta}(t)\right] = \int_0^\infty e^{-st} \cdot a\ddot{\theta}(t) \cdot dt = as^2\theta(s) - as\theta(to) - a\dot{\theta}(to), \tag{5.12a}$$

$$\text{(ii)} \quad L\left[b\dot{\theta}(t)\right] = \int_0^\infty e^{-st} \cdot b\dot{\theta}(t) \cdot dt = bs\theta(s) - b\theta(to), \tag{5.12b}$$

$$\text{(iii)} \quad L\left[c\theta(t)\right] = \int_0^\infty e^{-st} \cdot c\theta(t) \cdot dt = c\theta(s), \tag{5.12c}$$

$$\text{(iv)} \quad L\left[f(t)\right] = \int_0^\infty e^{-st} \cdot f(t) \cdot dt = F(s). \tag{5.12d}$$

where $\theta(s)$ denotes the Laplace transform of $\theta(t)$,

$F(s)$ denotes the Laplace transform of f(t),

$\theta(to)$ denotes the initial condition of $\theta(t)$ at time t = to,

and $\dot{\theta}(to)$ denotes the first derivative of $\theta(t)$ at time t = to.

Substituting the transforms of equations (5.12a) to (5.12d) into equation (5.11) anc collecting terms gives the following:

$$(as^2 + bs + c)\theta(s) = F(s) + (as + b)\theta(to) + a\dot{\theta}(to) \tag{5.13}$$

and rearranging:

$$\theta(s) = \frac{1}{(as^2 + bs + c)} \cdot F(s) + \frac{1}{(as^2 + bs + c)} ((as + b)\theta(to) + a\dot{\theta}(to)). \tag{5.14}$$

This transformation of the equation has two important effects:

(a) It has converted the differential equation into an algebraic equation, so that the solution variable θ can be separated explicitly, and

(b) It has separated the initial conditions, which appear explicitly in the second term on the RHS of equation (5.14). In general, for an equation of n-th degree, the first (n-1) initial derivatives are required.

We shall see that transforming the differential equation makes solving it a systematic and straightforward procedure, particularly for equations of degree 3 or higher.

Very often we are not interested in obtaining a solution with a specific set of initial conditions, and it is sufficient to find the *normal response*, which is that response obtained with zero initial conditions. If $\dot{\theta}(to) = \theta(to) = 0$ in equation (5.14), it reduces to the simple form:

$$\theta(s) = \frac{1}{(as^2 + bs + c)} \cdot F(s) \tag{5.15}$$

which is of the general form:

Response transform = Transfer Function x Input transform. (5.16)

5.3.1 Inverse Transformation

To find the response of a given system to an input function f(t) we multiply the Transfer Function of the system by the input transform $F(s)$. In accordance with equation (5.16) this results in the response transform $\theta(s)$, which is of course a function of s. All the information needed about the response can in principle be found from its transform, but it is often desirable to return to the time response $\theta(t)$. This process is known as inverse transformation.

There are three basic methods for finding the inverse of a transform:

(i) Inspection and comparison with a table of transforms - see Table 5.1. For example, a transform such as $F(s) = 1/(1 + 4s)$ can be arranged to $(\frac{1}{4})/(\frac{1}{4} + s)$ which is of the same form as No. 3 in Table 5.1. This gives the inverse immediately as $\frac{1}{4}\exp(-t/4)$.

TABLE 5.1 Laplace transform pairs

	F(s)	f(t) (t>0)	
1	1	$\delta(t)$	unit impulse
2	$\exp(-Ts)$	$\delta(t-T)$	delayed impulse
3	$1/(s + a)$	$\exp(-at)$	exponential decay
4	$1/(s + a)^n$	$(t^{n-1})\exp(-at))/(n-1)!$	$n = 1,2,3,...$
5	$1/s$	$u(t)$ or 1	unit step
6	$1/s^2$	t	unit ramp
7	$1/s^n$	$t^{n-1}/(n-1)!$	$(0! = 1)$ $n = 1,2,3...$
8	$\omega/(s^2 + \omega^2)$	$\sin(\omega t)$	
9	$s/(s^2 + \omega^2)$	$\cos(\omega t)$	
10	$1/(s + a)(s + b)$	$(e^{-at} - e^{-bt})/(b - a)$	$a \neq b$
11	$s/(s + a)(s + b)$	$(ae^{-at} - be^{-bt})/(a - b)$	$a \neq b$
12	$1/s(s + a)$	$(1 - e^{-at})/a$	
13	$1/(s^2 + 2\zeta\omega_n s + \omega_n^2)$	$(1/\omega d)\cdot(e^{-\zeta\omega_n t})\sin \omega dt$ $(\omega d = \omega_n \sqrt{1 - \zeta^2})$	$0 < \zeta < 1$
14	$s/(s + a)^2$	$(1 - at)\, e^{-at}$	
15	$a^2/(s + a)^2 s$	$1 - e^{-at}\,(1 + at)$	

(ii) Partial fraction expansion. This allows a complicated transform to be expressed as the sum of a number of simpler transforms.

For a transform of the form $(s+a)/(s+b)(s+c)^2$ there are three partial fractions, formed from the terms in the denominator:

$$\frac{(s + a)}{(s + b)(s + c)^2} \equiv \frac{A1}{(s + b)} + \frac{A2}{(s + c)} + \frac{A3s}{(s + c)^2} . \tag{5.17}$$

The coefficients A1, A2 and A3 can be found either by equating like powers of s in the identity or by substituting suitable values of s.

EXAMPLE 5.1. Find f(t) if F(s) =

$$\frac{(s + 1)}{(s + 3)(s + 2)^2} .$$

Expanding in partial fractions:

$$\frac{(s + 1)}{(s + 3)(s + 2)^2} = \frac{A1}{(s + 3)} + \frac{A2}{(s + 2)} + \frac{A3s}{(s + 2)^2}$$

Multiplying both sides by the LHS denominator:

$$(s + 1) = A1(s + 2)^2 + A2(s + 3)(s + 2) + A3s(s + 3) \tag{5.18}$$

expanding and collecting terms gives

$$s + 1 = (A1 + A2 + A3)s^2 + (4A1 + 5A2 + 3A3)s + (4A1 + 6A2)$$

equating like powers of s:

$$0 = A1 + A2 + A3,$$

$$1 = 4A1 + 5A2 + 3A3,$$

$$1 = 4A1 + 6A2$$

and solving these simultaneous equations gives

$$A1 = -2,\ A2 = 3/2,\ A3 = \tfrac{1}{2}.$$

The same results may be obtained by substituting s = 0, -2 and -3 successively in equation (5.18).

By either method, we have

$$F(s) = \frac{(s + 1)}{(s + 3)(s + 2)^2} = \frac{3}{2(s + 2)} + \frac{s}{2(s + 2)^2} - \frac{2}{(s + 3)}$$

The RHS terms correspond to entries 3 and 14 in Table 5.1 so the inverse is given by:

$$f(t) = (3/2)e^{-2t} + (\tfrac{1}{2})(1 - 2t)e^{-2t} - 2e^{-3t}.$$

(iii) The third method of inversion is to reduce a complicated transform to a set of simpler transforms by using Heaviside's Expansion Theorem, which gives a systematic procedure for finding the coefficients. It is stated as follows:

If we can express a function $F(s)$ as the ratio of numerator and denominator functions $N(s)$ and $D(s)$:

$$F(s) = \frac{N(s)}{D(s)} = \frac{N(s)}{(s - P1)(s - P2)\ldots(s - Pk)\ldots(s - Pq)^r} \qquad (5.19)$$

where $P1,\ldots,Pk$ are simple roots of $D(s)$

and Pq is a multiple root of multiplicity r

then the expansion is

$$F(s) = \frac{Ap1}{(s - p1)} + \frac{Ap2}{(s - p2)} + \ldots + \frac{Apk}{(s - pk)} +$$

$$+ \frac{Ap_q r}{(s - p_q)^r} + \frac{Ap_q(r-1)}{(s - p_q)^{r-1}} + \ldots + \frac{Ap_q 2}{(s - p_q)^2} + \frac{Ap_q 1}{(s - P_q)} \qquad (5.20)$$

The $P1$ to Pk and P_q may be real or complex, and the coefficients A are determined as follows:

For simple roots

$$Ap_n = \Big[(s - Pn)F(s)\Big]_{s = pn} = \left[\frac{(s - Pn)N(s)}{D(s)}\right]_{s = pn} \quad \text{for } n = 1 \text{ to } k \qquad (5.21)$$

and for multiple roots

$$Ap_q(r-n) = \frac{1}{n!} \cdot \frac{d^n}{ds^n}\Big[(s - P_q)^r F(s)\Big]_{s = P_q} \quad \text{for } n = 0, 1, 2\ldots,(r - 1) \qquad (5.22)$$

where $n!$ denotes factorial $n = n(n - 1)(n - 2) \ldots 3.2.1$

Note that $0!$ is defined as 1 and $1! = 1$.

Check. There is a simple check which can be applied to verify a calculation of the coefficients.

The *residues* are coefficients of terms in the expansion with first-degree denominators, that is, the coefficients $Ap1$, $Ap2$, $Ap3,\ldots$, Apk and $Ap_q 1$ in Equation (5.20) are all residues.

The degree of a polynomial in s is the highest power of s in the polynomial.

1. If the degree of $D(s)$ in equation (5.19) is one higher than that of $N(s)$, then

$$\Sigma \text{ (residues)} = 1.$$

2. If the degree of $D(s)$ is two or more higher than that of $N(s)$, then

$$\Sigma \text{ (residues)} = 0.$$

EXAMPLE 5.2. Find $f(t)$ if $F(s) = 1/(s + 1)(s + 2)(s + 3)^3$. Here we have two simple roots ($s = -1, -2$) and one multiple root ($s = -3$) of multiplicity 3.

From equation (5.20) the expansion of $F(s)$ is

$$\frac{1}{(s + 1)(s + 2)(s + 3)^3} = \frac{A1}{(s + 1)} + \frac{A2}{(s + 2)} + \frac{A33}{(s + 3)^3} + \frac{A32}{(s + 3)^2} + \frac{A31}{(s + 3)} .$$

The coefficients for simple roots are found from equation (5.21):

$$A1 = \left[\frac{\cancel{(s + 1)}}{\cancel{(s + 1)}(s + 2)(s + 3)^3}\right]_{s = -1} = 1/8,$$

$$A2 = \left[\frac{\cancel{(s + 2)}}{(s + 1)\cancel{(s + 2)}(s + 3)^3}\right]_{s = -2} = -1.$$

and for multiple roots from equation (5.22):

Here n in equation (5.22) takes on the values 0, 1 and 2.

$$A33 = \left[\frac{\cancel{(s + 3)^3}}{(s + 1(s + 2)\cancel{(s + 3)^3}}\right]_{s = -3} = 1/2 \; (n = 0),$$

$$A32 = \frac{1}{1!} \cdot \frac{d}{ds}\left[\frac{\cancel{(s + 3)^3}}{(s + 1)(s + 2)\cancel{(s + 3)^3}}\right]_{s = -3} \qquad (n = 1).$$

Note that cancellation and differentiation are carried out *before* the substitution.

$$A32 = \frac{d}{ds}\left[\frac{1}{(s + 1)(s + 2)}\right]_{s = -3} = \left[\frac{-2s - 3}{(s^2 + 3s + 2)^2}\right]_{s = -3}$$

$$= \underline{3/8.}$$

$$A31 = \frac{1}{2!}\frac{d^2}{ds^2}\left[\frac{1}{(s + 1)(s + 2)}\right]_{s = -3} \qquad (n = 2)$$

$$= \frac{1}{2}\frac{d}{ds}\left[\frac{-2s - 3}{(s^2 + 3s + 2)^2}\right]_{s = -3}$$

$$= \frac{1}{2}\left[\frac{6s^2 + 18 + 14}{(s^2 + 3s + 2)^3}\right]_{s = -3}$$

$$= \underline{7/8.}$$

Applying the check mentioned above, in this case the degree of $D(s)$ is 5, the degree of $N(s)$ is zero, so we expect Σ (residues) = 0.
The residues are A1, A2 and A31:

$$1/8 - 1 + 7/8 = 0 \quad \underline{\text{check}},$$

The expansion of $F(s)$ is now explicitly:

$$F(s) = \frac{1}{8(s+1)} - \frac{1}{(s+2)} + \frac{1}{2(s+3)^3}$$

$$+ \frac{3}{8(s+3)^2} + \frac{7}{8(s+3)} .$$

Each of these terms is of a form which appears in Table 5.1, and by comparison f(t) can be obtained:

$$f(t) = 0.125 \exp(-t) - \exp(-2t)$$

$$+ (0.5t^2 + t + 0.875) \exp(-3t).$$

Note that setting $t = 0$ in the above equation gives an initial value of zero.

5.3.2 Complex Zeros

Occasionally a quadratic term occurs in the denominator which cannot be factorised into two real roots: the roots are complex. If the coefficients of the quadratic term are real, then the two roots will always occur as a complex conjugate pair.

For a function

$$F(s) = \frac{N(s)}{D(s)} = \frac{N(s)}{(s^2 + As + B)(s - r_1)\dots(s - r_m)}$$

with one quadratic root and m simple roots in the denominator, we have the partial fraction expansion.

$$F(s) = \frac{K_p}{s - a - jb} + \frac{K_n}{s - a + jb} + \frac{K_1}{s - r_1} + \dots + \frac{K_m}{s - r_m}$$

where $(s - a - jb)(s - a + jb) = s^2 + As + B$

and $j = \sqrt{-1}$.

The inverse transform is

$$f(t) = K_p \exp(a + jb)t + K_n \exp(a - jb)t$$

$$+ K_1 \exp r_1 t + \dots + K_m \exp r_m t$$

from entry no. 3 in Table 5.1. The coefficients K_p and K_n can be obtained from Heavisides expansion theorem in the usual way:

$$K_p = \left[\frac{(s - a - jb)\ N(s)}{(s^2 + As + B)(s - r_1)\ \dots\ (s - r_m)} \right]_{s = a + jb}$$

$$= \left[\frac{N(s)}{(s - a + jb)(s - r_1)\dots(s - r_m)}\right]_{s = a + jb}$$

$$= K(a + jb)/j2b$$

where
$$K(a + jb) = \left[(s^2 - 2as + a^2 + b^2)\ \frac{N(s)}{D(s)}\right]_{= a + jb}$$

Similarly

$$K_n = \left[\frac{(s - a + jb)\ N(s)}{(s^2 + As + B)(s - r_1)\dots(s - r_m)}\right]_{s = a - jb}$$

$$= K(a - jb)/(- j2b).$$

The constants K(a + jb) and K(a - jb) are complex conjugate:

$$K(a + jb) = |K(a + jb)|e^{j\alpha}$$

$$K(a - jb) = |K(a + jb)|e^{-j\alpha}$$

where
$$\alpha = \underline{/K(a + jb)}.$$

Hence
$$K_p = (1/j2b)\ |K(a + jb)|\ e^{j\alpha}$$

and
$$K_n = (-1/j2b)|K(a + jb)|e^{-j\alpha}$$

so that the inverse function is given by

$$f(t) = (1/b)\ |K(a + jb)\ |e^{at}\left[\frac{e^{j(bt+\alpha)} - e^{-j(bt+\alpha)}}{2j}\right]$$

$$+ K_1\ e^{r_1 t} + \dots + K_m e^{r_m t}$$

$$= (1/b)|\ K(a + jb)\ |e^{at}\ \sin(bt + \alpha) + k_1 e^{r_1 t} + \dots + K_m\ d^{r_m t}$$

since the expression in square brackets is of the form $(e^{jx} - e^{-jx})/j2$ which is a definition of sin x.

<u>EXAMPLE 5.3</u>. Find the inverse transformation of the function

$$\frac{40}{(s^2 + 2s + 5)(s + 3)}.$$

For the quadratic term in the denominator

$$s^2 + 2s + 5 = (s - a - jb)(s - a + jb)$$

$$= s^2 - 2as + a^2 + b^2,$$

equating coefficients we obtain

$$a = -1,\ b = 2$$

$$K(a + jb) = \left[\frac{40}{s + 3}\right]_{s = -1 + j2} = 10 - j10$$

$$|K(a + jb)| = \sqrt{100 + 100} = 14.14,$$

$$\alpha = \angle K(a + jb) = \tan^{-1}(-10/10) = -\pi/4,$$

$$K_1 = \left[\frac{40}{s^2 + 2s + 5}\right]_{s = -3}$$

The pulse response is thus

$$f(t) = (14.14/2)e^{-t}\sin(2t - \pi/4) + 5e^{-3t}$$

$$= 7.07\ e^{-t}\sin(2t - 0.785) + 5e^{-3t}$$

which is sketched in Fig. 5.6.

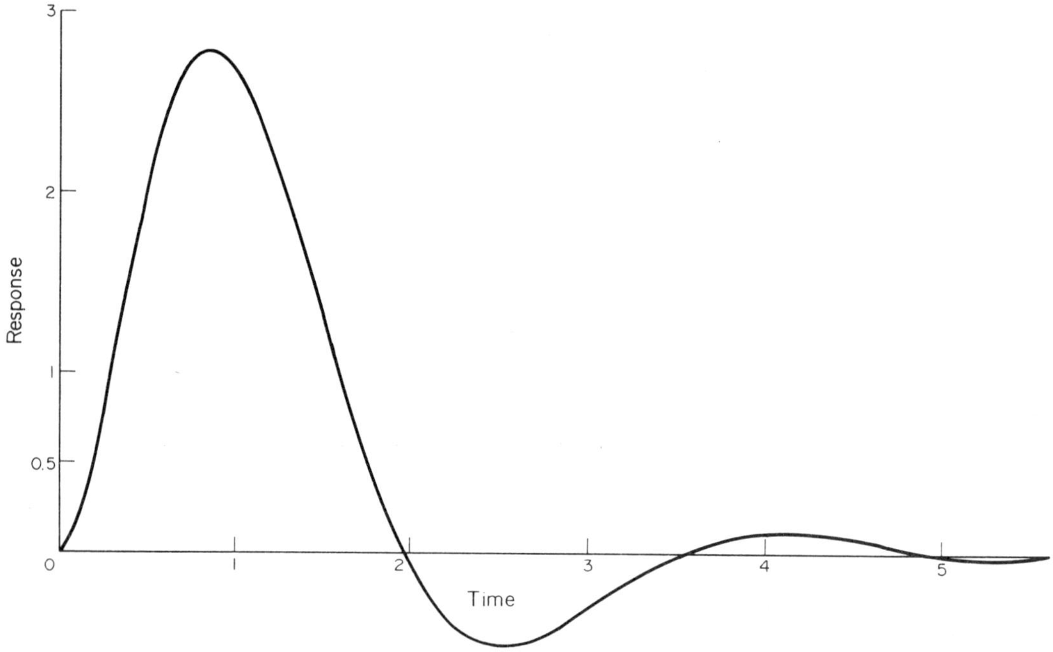

Fig. 5.6. Pulse response of transfer function $\frac{40}{(s^2 + 2s + 5)(s + 3)}$

5.3.3 Transform Theorems

Four useful theorems, in two complementary pairs, are stated here without proof. Proofs are available in many standard texts, such as the one by Healey.

$L(f(t)) = F(s)$ in each case.

1a. Initial value theorem:

$$\lim_{t \to 0} f(t) = \lim_{s \to \infty} sF(s).$$

Stated in words, the initial value of f(t) (the value at zero time) is found by multiplying its transform, F(s), by s and finding the limit as s approaches infinity.

1b. Final value theorem:

$$\lim_{t \to \infty} f(t) = \lim_{s \to 0} sF(s).$$

This is the dual of the initial value theorem: the final value of f(t) (the value after a long time) is found by multiplying its transform, F(s), by s and taking the limit as s approaches zero.

2a. Translation in time (shift theorem):

$$L(f(t - T)) = e^{-Ts} F(s) \text{ where T is real.}$$

If a function f(t) is subjected to a time delay, T, the transform of the delayed function is obtained by multiplying the transform of the original function by e^{-Ts}.

2b. Translation in the s-domain:

$$L(e^{at}f(t)) = F(s - a) \text{ where a is real or complex.}$$

If a function f(t) has the transform F(s), then the transform of $e^{at}f(t)$ can be found by replacing s in F(s) by s - a.

5.4 APPLICATION OF LAPLACE TRANSFORMS TO THE ANALYSIS OF A HEAT EXCHANGER

In order to illustrate some of the techniques introduced in the previous sections, they will be applied to the analysis of a simple heat exchanger. The time response of the exchanger to various inputs can then be predicted, as well as the effect of adding a controller.

The heat exchanger is illustrated in Fig. 5.7. The primary mean temperature is θ_p; secondary water enters at a temperature θ_1 and is heated by the tubes to θ_2. In order to avoid the complications of temperature gradients within the shell of the exchanger, the simple assumption will be made that the secondary water is immediately heated to its leaving temperature θ_2 as soon as it enters the shell. The rate of flow of water through the secondary is q kg/s. The volume of water contained in the secondary is V m^3 and the specific heat capacity of the water is C_1 J/kg°C, with density ρ kg/m^3. The total exposed surface area of the tubes is A_1 m^2 and the heat-transfer coefficients at the exposed surface is h_1 W/m^2°C. It may be assumed that this external coefficient is the one which controls the overall heat transfer to the water.

Writing a heat balance on the secondary water, we have

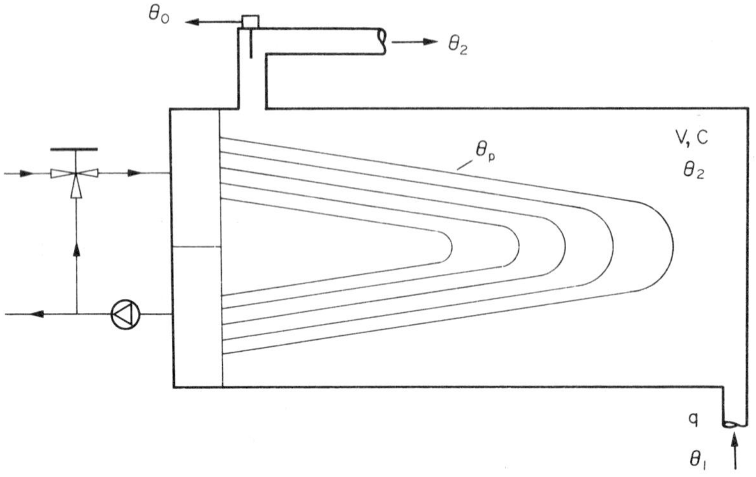

Fig. 5.7. Heat exchanger.

Rate of heat supply from primary = Rate of heat loss from secondary + Rate of heat storage, in secondary

and in symbols

$$(\theta_p - \theta_2)h_1A_1 = qC_1\,(\theta_2 - \theta_1) + \rho VC_1(d\theta_2/dt). \tag{5.23}$$

Rearranging and transforming, with zero initial conditions, this gives

$$\theta_2(s) = \left[\frac{1}{1 + qC_1/h_1A_1 + (\rho VC_1/h_1A_1)s}\right]\theta_p\,(s) + \left[\frac{qC_1/h_1A_1}{1 + qC_1/h_1A_1 + (\rho VC_1/h_1A_1)s}\right]\theta_1\,(s) \tag{5.24}$$

showing how θ_2 is influenced by θ_p and by θ_1 separately via the two transfer functions in square brackets.

Examining the units of the two groups of symbols (qC_1/h_1A_1) and $(\rho VC_1/h_1A_1)$ shows that the first group is dimensionless whereas the second has the dimensions of time. Equation (5.24) may be rewritten as

$$\theta_2(s) = \left[\frac{1}{1 + \alpha + Ts}\right]\theta_p(s) + \left[\frac{\alpha}{1 + \alpha + Ts}\right]\theta_1(s) \tag{5.25}$$

where $\alpha = qC_1/h_1A_1$ and $T = \rho VC_1/h_1A_1$.

Dividing the numerator and denominator of both transfer functions by $1 + \alpha$ we obtain the simplest form:

$$\theta_2(s) = \frac{1}{1 + \alpha}\left[\frac{1}{1 + T's}\right]\theta_p(s) + \frac{\alpha}{1 + \alpha}\left[\frac{1}{1 + T's}\right]\theta_1(s) \tag{5.26}$$

where $T' = T(1 + \alpha)$.

This shows that the form of the dynamic response of the water-outlet temperature is the same for changes in each input, θ_p and θ_1, because the term in square brackets is the same for each. The size of the response, however, is determined by the multipliers $1/(1 + \alpha)$ and $\alpha/(1 + \alpha)$. The value of α depends mainly on the flow rate of secondary water and it appears that for a large flow rate ($\alpha >> 1$) the change in θ_2 due to a unit change in θ_p will be small, whereas the change in θ_2 produced by a unit change in θ_1 will be nearly unity.

A block diagram representation of equation (5.26) is shown in Fig. 5.8, first in the direct form (a) as given by the equation, then in a simplified form (b).

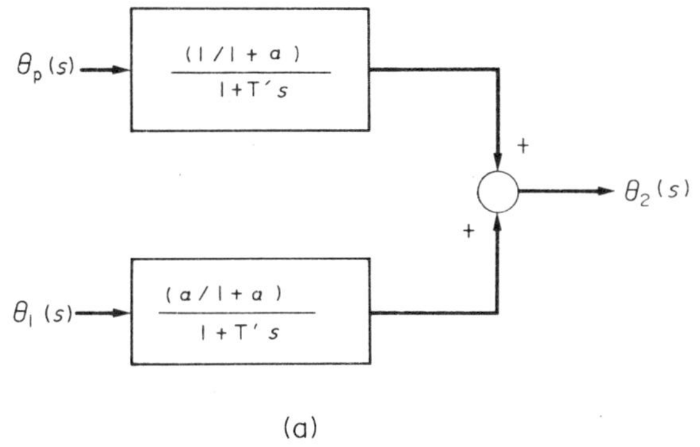

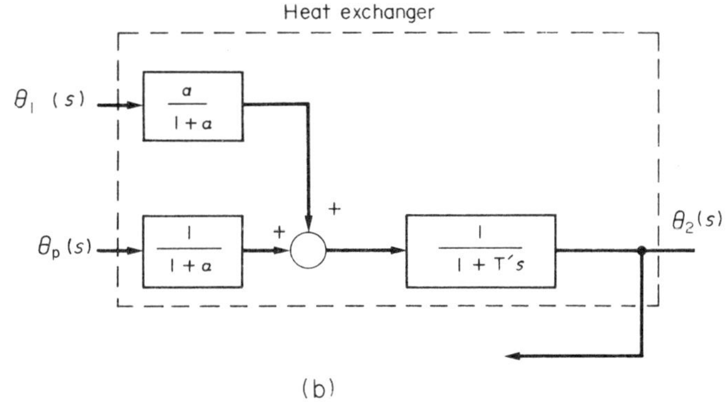

Fig. 5.8. Heat exchanger block diagrams.

5.4.1 Heat-exchanger Response - Without Controller

Before including a controller in the system, let us calculate the response of the outlet temperature of the secondary water to a step change of 10°C magnitude in the inlet water temperature. If the primary temperature θ_p is assumed to be constant, then $\theta_p(s)$ can be taken as zero, and $\theta_2(s)$ will represent the *change* in the outlet temperature of the secondary water.

If $\theta_p(s) = 0$, we have, from equation (5.26):

$$\theta_2(s) = 0 + \frac{\alpha}{(1+\alpha)}\left[\frac{1}{1+T's}\right]\theta_1(s).$$

A unit step change has the transform $1/s$ (see entry 5 in Table 5.1), so the transform of a step change of magnitude $10^{\circ}C$ will be $10/s$.

Hence $\theta_1(s) = 10/s$ and we can take values of $\alpha = 9$ and $T' = 100$ seconds.

Substituting these values gives

$$\theta_2(s) = \frac{9}{10}\left[\frac{1}{(1+100s)}\right]\cdot\frac{10}{s} = \frac{9}{(1+100s)s} = \frac{0.09}{(0.01+s)s}\,. \qquad (5.27)$$

The time response of θ_2 can be found by carrying out the inversion procedure on this transform. It is interesting, however, first to apply the final value theorem (see Section 5.3.3) to find the rise in temperature that will occur after a long time.

$$\text{Final value} = \lim_{s\to 0}\frac{0.09}{(0.01+s)} = 9.$$

This indicates that eventually the secondary water outlet temperature will rise by $9^{\circ}C$ in response to the $10^{\circ}C$ rise in inlet temperature.

In order to find the time response $\theta_2(t)$ we first expand $\theta_2(s)$ in partial fractions:

$$\theta_2(s) = \frac{0.09}{(0.01+s)s} = \frac{Ao}{s} + \frac{A0.01}{(0.01+s)}\,.$$

Heaviside's expansion theorem may then be used to determine the coefficients (see Section 5.3.1).

$$Ao = \left[\frac{0.09}{(0.01+s)}\right]_{s=0} = 9,$$

$$A0.01 = \left[\frac{0.09}{s}\right]_{s=0.01} = -9.$$

Applying the check mentioned after Example 5.1, here we have two residues and the degree of the denominator of $\theta_2(s)$ is two higher than that of the numerator, so we expect that

$$\Sigma\ (\text{residues}) = 0$$

and this is in fact the case, since

$$9 - 9 = 0.$$

The fact that this agreement is obtained does not, of course, guarantee that the coefficients are correct, but a failure of this test would indicate that at least one of them is incorrect.

Now substituting these values into the partial fraction expansion of $\theta_2(s)$, we obtain:

$$\theta_2(s) = 9/s - 9/(0.01+s).$$

These partial fractions correspond to entries 3 and 5 in Table 5.1, so the inversion is straightforward:

$$\theta_2(t) = 9 - 9\ \exp(-0.01t).$$

This is shown in Fig. 5.9. It is apparent that the result obtained above from the final value theorem is confirmed. It should be noted that Fig. 5.9 illustrates the response to a disturbance input in the absence of any feedback control action. In the next section we shall observe the effect of adding a controller.

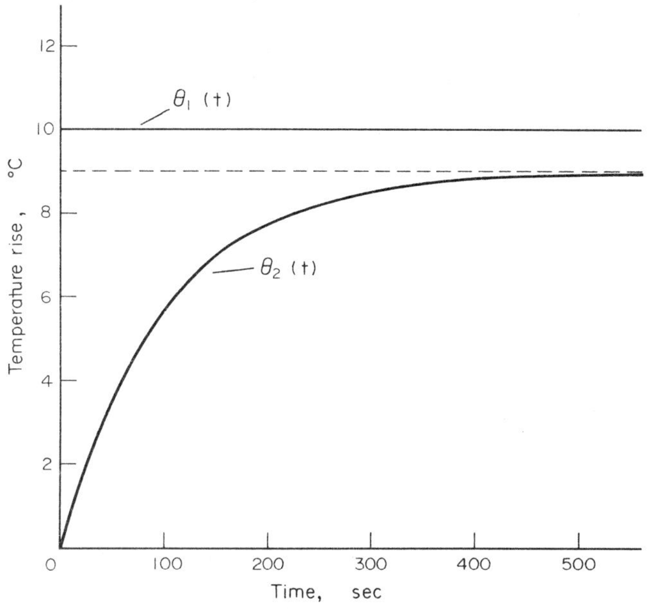

Fig. 5.9. Step response without controller.

5.4.2 Heat-exchanger Response - With Controller

Rearranging the block diagram of Fig. 5.8(b) and adding a controller to correspond to the diagram of Fig. 5.2, we obtain the arrangement shown in Fig. 5.10.

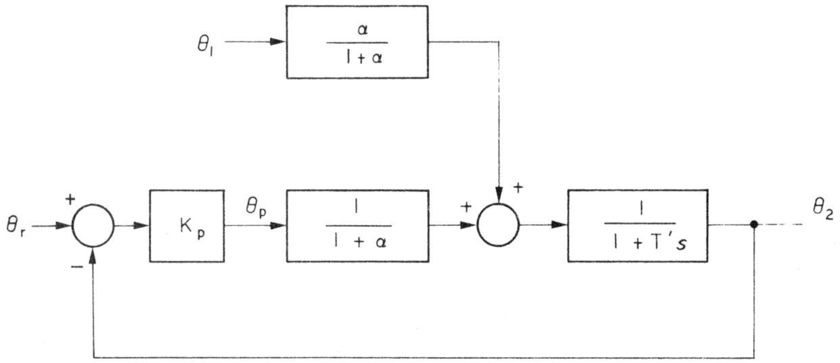

Fig. 5.10. Block diagram for heat exchanger with controller.

Analysing the block diagram of Fig. 5.10 gives the relationship between the two input variables θ_1 (the secondary water-inlet temperature and θ_r (the reference input or set-point of the controller) and the output variable θ_2 (the secondary water-outlet temperature):

$$\theta_2(s) = \frac{\alpha}{(1+\alpha)}\left[\frac{1}{1+K'+Ts}\right]\theta_1(s) + K'\left[\frac{1}{1+K'+Ts}\right]\theta_r(s) \qquad (5.28)$$

Where $K' = \dfrac{K}{(1+\alpha)}$.

In this case a proportional controller has been used since this is analytically the simplest type. It is represented by a simple proportional gain K. The value of K is normally under the control of the user as a setting on the front panel of the instrument, usually scaled in terms of the proportional band. Small values of proportional band correspond to large values of K. The two transfer functions in equation (5.28) represent the responses to the two input variables. In order to compare the response of the controlled system with that of the system without the controller as given in the previous section, we can consider a step change in θ_1 of magnitude 10°C as before, and zero change in θ_r. The effective time constant T of the exchanger can be taken as 100 seconds as before and α as 9. Substituting into equation (5.28) gives

$$\theta_2(s) = \frac{9}{10}\;\frac{1}{1+K'+100s}\;\frac{10}{s} + 0$$

$$= \frac{9}{(1+K'+100s)s}. \qquad (5.29)$$

Evidently for $K' = 0$ this reduces to equation (5.27), the response in the absence of a controller. We can now examine the response for various non-zero values of the proportional gain K.

First obtaining the general result for $\theta_2(t)$, we expand equation (5.29) in partial fractions:

$$\theta_2(s) = \frac{0.09}{\left(\frac{1+K'}{100}+s\right)s} = \frac{A}{s} + \frac{B}{\left(\frac{1+K'}{100}+s\right)}.$$

By Heaviside's theorem:

$$A = \left[\frac{0.09}{\frac{1+K'}{100}+s}\right]_{s=0} = \frac{9}{1+K'}$$

$$B = \left[\frac{0.09}{s}\right]_{s = -\frac{(1+K')}{100}} = \frac{-9}{1+K'}.$$

Hence $$\theta_2(t) = \frac{9}{1+K'}\left[1 - \exp(-t/T')\right] \qquad (5.30)$$

where $$T' = \frac{100}{1+K'} \text{ and } K' = \frac{K}{1+\alpha}.$$

Examining equation (5.30) and the definition of the time-constant T', it appears that an increase in the proportional gain K will have two important effects on the response:

(i) The multiplying factor $9/1 + K'$ will be reduced, so the size of the response to the disturbance input θ_1 will decrease; and

(ii) The time-constant T will be reduced, so the speed of response will be improved.

The response from equation (5.30) is plotted in Fig. 5.11 for various values of gain K. It can be seen that the larger the value of K, the smaller is the

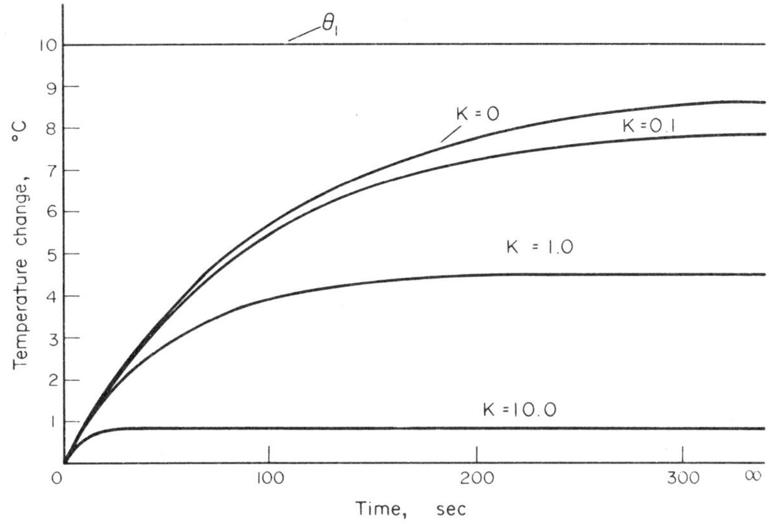

Fig. 5.11. Responses to disturbance input for various values of proportional gain.

response. This is to be expected if one recalls that θ_1 is a disturbance to the system and an ideal or perfect controller would respond so that no change occurred in θ_2 whatever the change in θ_1. This simple system produces the theoretical result that increasing the gain K to infinity will produce an ideal response. In practice no system is ever quite so simple as this: further time-constants and time delays always exist and these produce a response which is initially oscillatory as K increases, and eventually is unstable.

Returning to equation (5.28) we can also examine the effect of K on the response to changes in θ_r, the reference input variable. As before, we take a 10°C step change so that $\theta_r(s) = 10/s$ and $\theta_1(s) = 0$.

Hence

$$\theta_2(s) = 0 + K' \left[\frac{1}{1 + K' + 100s}\right] \frac{10}{s} .$$

Expanding in partial fractions as before and inverting gives the general result:

$$\theta_2(t) = \frac{10K'}{1 + K'} \left[1 - \exp(-t/T')\right] \tag{5.31}$$

where $T' = \frac{100}{1 + K'}$ and $K' = \frac{K}{1 + \alpha}$ as before.

Comparing this result with equation (5.30) we see that the dynamic term within the square brackets is the same as before, so that large values of K will give fast response, but the multiplying factor now approaches 10 as K increases. This is shown in Fig. 5.12 for various values of K. Again as K approaches infinity the

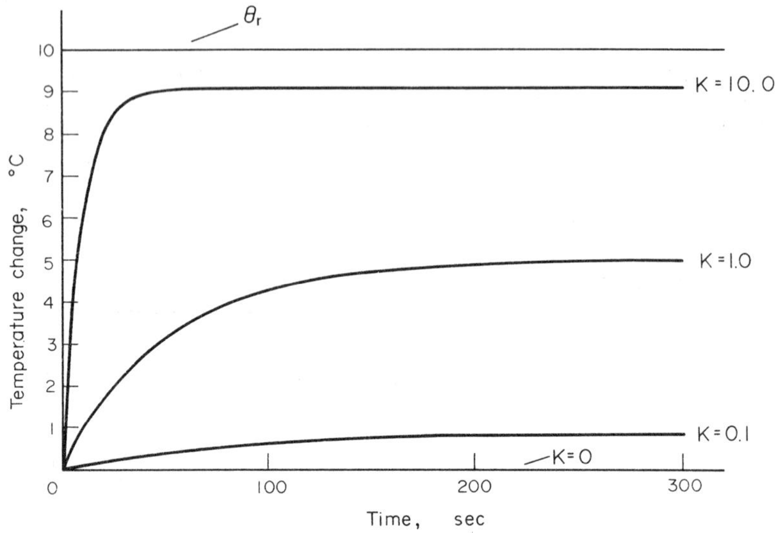

Fig. 5.12. Response to reference input for various values of proportional gain.

response becomes more perfect since we are here considering a change in the reference input. An ideal controller should give a response in the output which matches θ_r precisely. For this example the larger the value of K the better is the controlled performance. The steady-state difference between the reference input θ_r and the actual value of the controlled variable θ_2 is known as the OFFSET. For a perfect controller the offset should be zero, and this condition would be approached with large values of proportional gain.

5.5 FIRST-ORDER STEP RESPONSES

The type of response curve seen in Figs. 5.11 and 5.12 is described by a general equation of the form:

$$\theta(t) = \theta_F \, (1 - \exp\,(-t/T)) \qquad (5.32)$$

where $\theta(t)$ = response at time t,

θ_F = final value of response

and T = the time constant of the response.

The responses of a large number of processes and components can be characterised by a simple equation of this type: thermometer bulbs, mixing processes, ventilation effects, acoustic reverberation, etc. It is worthwhile examining some of its properties in more detail. This response curve is often obtained experimentally and the value of T has to be found from the observed data. There are a number of methods of doing this which are listed below.

(a) *63% rise time*

If we substitute t = T in equation (5.32) above, we obtain

$$\theta(T) = \theta_F\ (1 - \exp(-1))$$

$$= 0.632\theta_F.$$

Hence the time constant can be determined as the time taken for the response to change by 63% of its final change, (see Fig. 5.13). This definition is sometimes applied in more complex systems which it may be convenient to describe as if they were first order.

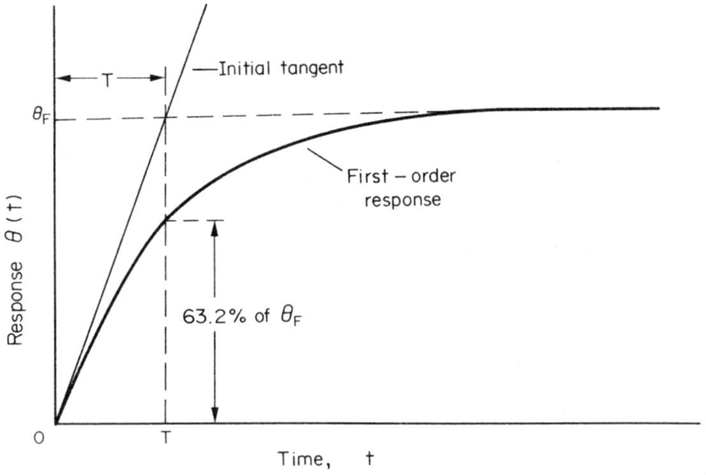

Fig. 5.13. Finding the time constant from the first-order step response - initial slope.

(b) *Initial slope*

Differentiating equation (5.32) gives:

$$\frac{d\theta(t)}{dt} = (\theta_F/T)\ \exp\ (-t/T)$$

and at zero time this gives a slope of θ_F/T. This is the slope of a tangent to the curve drawn through the origin, as in Fig. 5.13. Hence the point where the tangent cuts the final value θ_F will be at a time T after the start.

(c) *Any tangent*

It is not often realised that the tangent at any point on the curve can be used to find a value of T. Referring to Fig. 5.14, the equation of the straight line AB is

$$\theta = \left(\frac{d\theta}{dt}\right)_{t_1} \cdot t + C$$

where C is the intercept on the vertical axis.

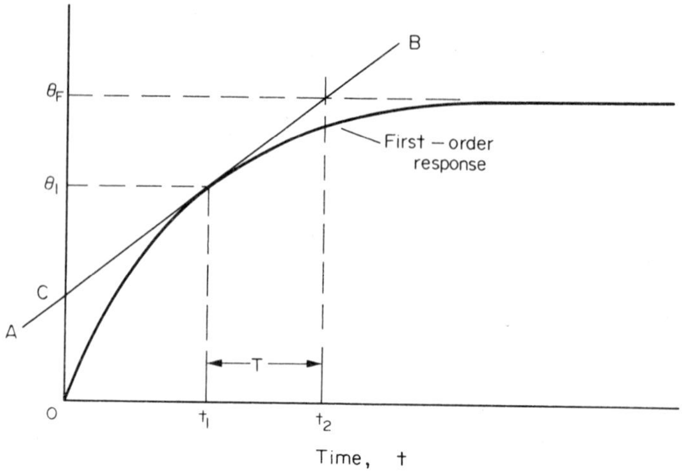

Fig. 5.14. Finding the time constant from the first-order step response - any tangent.

From the results above we have

$$\frac{d\theta}{dt} = \frac{\theta_F}{T} \exp(-t/T)$$

so that

$$\left(\frac{d\theta}{dt}\right)_{t_1} = \frac{\theta_F}{T} \exp(-t_1/T).$$

Substituting into the equation of the straight line gives

$$\theta = t \cdot \frac{\theta_F}{T} \exp(-t_1/T) + C$$

and rearranging:

$$C = \theta - t \cdot (\theta_F/T)\exp(-t_1/T)$$

which is constant for that straight line.

Denoting the time at which the tangent reaches θ_F by t_2, we can write

$$C = \theta_F - t_2 (\theta_F/T)\exp(-t_1/T)$$

and substituting this back into the equation for the straight line gives

$$\theta = t \cdot \frac{\theta_F}{T} \exp(-t_1/T) + \theta_F - t_2 \frac{\theta_F}{T} \exp(-t_1/T),$$

when $t = t_1$ and $\theta = \theta_1$ this becomes

$$\theta_1 = \theta_F - (t_2 - t_1)(\theta_F/T)\exp(-t_1/T)$$

or

$$\theta(t_1) = \theta_F \left(1 - \frac{(t_2 - t_1)}{T} \exp(-t_1/T)\right)$$

and comparing this equation with equation (5.32) it is apparent that

$$t_2 - t_1 = T.$$

This shows that it is possible to obtain a number of estimates of a first-order time constant by drawing various tangents to the curve.

(d) *Transformation to a straight line*

The results obtained above use either a single point on the response curve or a slope which cannot usually be determined with any precision. It seems preferable to use a method which takes into account most or all of the experimental points. This can be done by transforming the response curve so that the best straight line can be chosen to pass through the whole set of measured points. A least-squares regression calculation could be used to give the best line in this case.

Returning to equation (5.32) and rearranging, we can write:

$$\theta(t)/\theta_F = 1 - \exp(-t/T)$$

and

$$\frac{\theta_F - \theta(t)}{\theta_F} = \exp(-t/T),$$

$$\frac{\theta_F}{\theta_F - \theta(t)} = \exp(t/T).$$

Taking the natural logarithm of each side:

$$\ln\left[\frac{\theta_F}{\theta_F - \theta(t)}\right] = t/T$$

and
$$T \cdot \ln\left[\frac{\theta_F}{\theta_F - \theta(t)}\right] = t.$$

This is the equation of a straight line passing through the origin, with slope 1/T, as in Fig. 5.15. Taking pairs of observed values of t and $\theta(t)$ gives a plot of points which can be used to estimate a best straight line, the slope of which gives a measure of T.

5.5.1 Time Constants of Simple Thermometer Bulbs

It was shown in Section 5.2 that the time constant T of a thermometer bulb could be expressed as the product of its thermal capacity and the total thermal resistance at its exposed surface.

$$T = \rho CV/hA.$$

The heat-transfer coefficient h and hence the time constant are very strongly

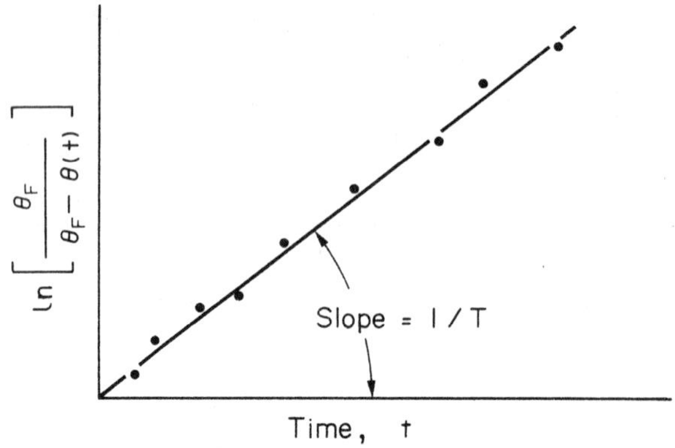

Fig. 5.15. Finding T by transformation to a straight line.

dependent on the properties of the fluid adjacent to the surface. The usual fluids of interest are air and water. With forced convection the value of h increases rapidly with velocity. Hence the time constant decreases rapidly as the velocity goes up, as shown in Fig. 5.16. From the heat-transfer relationships, it can be shown that for a given size and construction of sensor the ratio of T for water and air flowing at similar speeds is about 300:1.

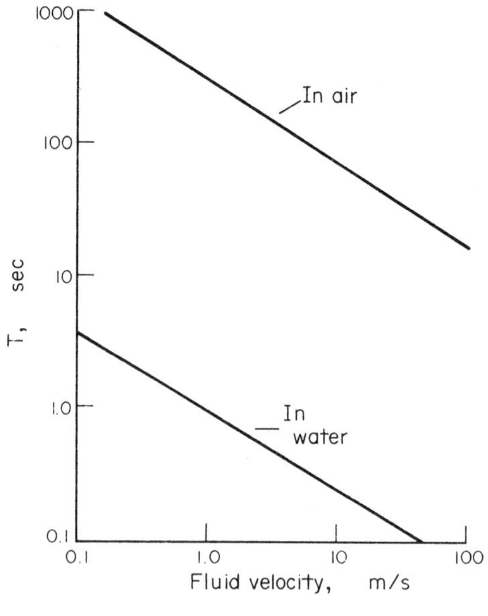

Fig. 5.16. The effect of fluid velocity on the thermal time constant of a sensor.

5.6 SECOND-ORDER STEP RESPONSE

The behaviour of the first-order system outlined in Section 5.4 illustrated the effects of proportional control. To show how an oscillatory step response can occur we now examine a simple second-order system, as shown in Fig. 5.17. The two blocks with time constants T_1 and T_2 could be regarded as representing the primary

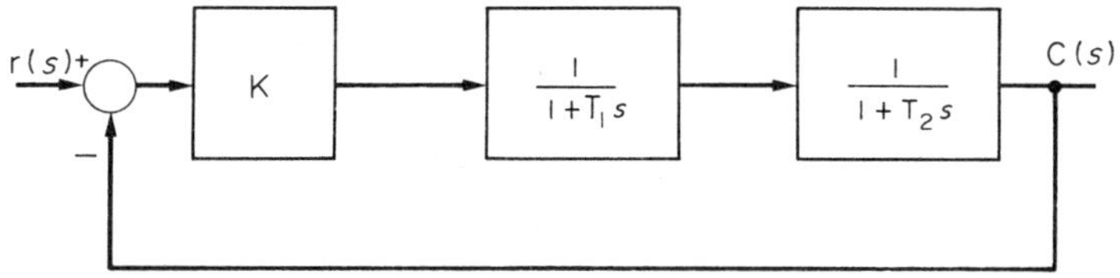

Fig. 5.17. A simple second-order system.

tubes and the secondary water shell of the heat exchanger of Fig. 5.7.

Analysing the block diagram of Fig. 5.17 we can write down the overall transfer function connecting the controlled variable $c(s)$ and the reference variable $r(s)$:

$$c(s) = \left[\frac{K}{1 + K + (T_1 + T_2)s + T_1T_2s^2}\right] r(s)$$

$$= \frac{K}{1+K}\left[\frac{1}{1 + \left(\frac{T_1 + T_2}{1 + K}\right)s + \left(\frac{T_1T_2}{1 + K}\right)s^2}\right] r(s). \qquad (5.33)$$

The analysis of second-order systems of the same general form as that in equation (5.33) occurs in a number of branches of engineering and it is customary to express the function in square brackets in terms of a standard form:

$$c(s) = \frac{K}{1+K}\left[\frac{1}{1 + \frac{2\zeta}{\omega_h}s + \frac{1}{\omega_n^2}s^2}\right] r(s)$$

$$= \frac{K}{1+K}\left[\frac{\omega_n^2}{\omega_n^2 + 2\zeta\omega_n s + s^2}\right] r(s) \qquad (5.34)$$

The two new variables in this form are ζ(ZETA), the damping ratio, a dimensionless real positive number, and ω_n (OMEGA), the undamped natural frequency, radians/second. They are related to the original variables by comparing equations (5.33) and (5.34):

$$\zeta = \frac{T_1 + T_2}{2\sqrt{T_1T_2}} \cdot \frac{1}{\sqrt{1 + K}}, \qquad (5.34a)$$

and

$$\omega_n = \sqrt{\frac{1 + K}{T_1T_2}}. \qquad (5.34b)$$

It can be seen that increasing the value of K produces a decrease in ζ (less damping, faster response) and an increase in the value of ω_n.

In order to examine the dynamic behaviour of this system is will be convenient first of all to ignore the effect of the multiplying factor $K/(1 + K)$ and to consider only the transfer function in square brackets in equation (5.34). Afterwards the two effects can be combined (Section 5.6.2).

5.6.1 Dynamic Response of Second-order Transfer Function

To obtain the response of the transfer function to a step change of 10°C magnitude in the reference input r, we set $r(s) = 10/s$:

$$c(s) = \frac{10\,\omega_n^2}{(\omega_n^2 + \zeta 2\omega_n s + s^2)s} \tag{5.35}$$

$$= \frac{10\omega_n^2}{[s + \omega_n(\zeta - \alpha)][s + \omega_n(\zeta + \alpha)]s}$$

by factorising

where $\alpha = \sqrt{\zeta^2 - 1}$.

Expressing this in partial fractions:

$$c(s) = \frac{A}{s} + \frac{B}{s + \omega_n(\zeta + \alpha)} + \frac{C}{s + \omega_n(\zeta - \alpha)}$$

and from Heaviside's theorem the coefficients are

$$A = 10,\ B = \frac{10}{2\alpha(\zeta + \alpha)},\ C = \frac{-10}{2\alpha(\zeta - \alpha)}$$

so that

$$c(s) = \frac{10}{s} + \frac{10}{2\alpha(\zeta + \alpha)[s + \omega_n(\zeta + \alpha)]} - \frac{10}{2\alpha(\zeta - \alpha)[s + \omega_n(\zeta - \alpha)]}.$$

Again, these terms are of the form of entries 3 and 5 in Table 5.1 so that the inversion is straightforward:

$$c(t) = 10 + \frac{10}{2\alpha}\left[\frac{\exp(-\omega_n(\zeta + \alpha)t)}{(\zeta + \alpha)} - \frac{\exp(-\omega_n(\zeta - \alpha)t)}{(\zeta - \alpha)}\right]$$

$$= 10 + \frac{10e^{-\zeta\omega_n t}}{2\alpha}\left[\frac{(\zeta - \alpha)\,e^{-\alpha\omega_n t} - (\zeta + \alpha)e^{\alpha\omega_n t}}{\zeta^2 - \alpha^2}\right]$$

and $\zeta^2 - \alpha^2 = 1$.

Rearranging, we have

$$c(t) = 10 + \frac{10e^{-\zeta\omega_n t}}{2\alpha}\left[\zeta(e^{-\alpha\omega_n t} - e^{\alpha\omega_n t}) - \alpha(e^{-\alpha\omega_n t} + e^{\alpha\omega_n t})\right]$$

$$= 10 - \frac{10e^{-\zeta\omega_n t}}{2\alpha}\left[\zeta(e^{\alpha\omega_n t} - e^{-\alpha\omega_n t}) + \alpha(e^{\alpha\omega_n t} + e^{-\alpha\omega_n t})\right]$$

$$= 10 - 10e^{-\zeta\omega_n t}\left[\frac{\zeta}{\alpha}\frac{(e^{\alpha\omega_n t} - e^{-\alpha\omega_n t})}{2} + \frac{(e^{\alpha\omega_n t} + e^{-\alpha\omega_n t})}{2}\right].$$

Writing this out in full gives the general form of step-response equation:

$$c(t) = 10 - 10e^{-\zeta\omega_n t}\left[\frac{\zeta}{\sqrt{\zeta^2 - 1}}\frac{(\exp(\omega_n\sqrt{\zeta^2 - 1})t - \exp(-\omega_n\sqrt{\zeta^2 - 1})t}{2}\right.$$

$$\left. + \left(\frac{\exp(\omega_n\sqrt{\zeta^2 - 1})t + \exp(-\omega_n\sqrt{\zeta^2 - 1})t}{2}\right)\right] \tag{5.36}$$

This is valid for any value of ζ, except unity, for which case equation (5.35) reduces to the simpler form:

$$c(s) = \frac{10\omega_n^2}{(s + \omega_n)^2 s}. \tag{5.37}$$

Taking various ranges of values of ζ we can use equation (5.36) to observe the effect on the step response

(a) $\zeta = 0$ - *zero damping*

In this case

$$\sqrt{\zeta^2 - 1} = \sqrt{-1} = j$$

and equation (5.36) reduces to

$$c(t) = 10 - 10\left(\frac{\exp(j\omega_n t) + \exp(-j\omega_n t)}{2}\right)$$

and since $\cos x = (e^{jx} + e^{-jx})/2$ this becomes

$c(t) = 10 - 10 \cos \omega_n t$ which is shown in Fig. 5.18.

(b) $0 < \zeta < 1$ - *underdamped case*

In this case

$$\zeta^2 < 1, \text{ so } \sqrt{\zeta^2 - 1} = j\sqrt{1 - \zeta^2}.$$

The response now becomes

$$c(t) = 10 - 10e^{-\zeta\omega_n t}\left[\frac{\zeta}{\sqrt{1 - \zeta^2}} \sin \omega_n \sqrt{1 - \zeta^2}t\right.$$

$$\left. + \cos \omega_n \sqrt{1 - \zeta^2}t\right]$$

$$= 10 - 10e^{-\zeta\omega_n t}\left[\frac{\zeta}{\sqrt{1-\zeta^2}}\sin\omega_d t + \cos\omega_d t\right]$$

$$= 10 - \frac{10e^{-\zeta\omega_n t}}{\sqrt{1-\zeta^2}}\sin(\omega_d t + \phi)$$

where $\omega_d = \omega_n\sqrt{1-\zeta^2}$ and $\cos\phi = \zeta$.

This is basically similar to the previous case, except that the sine wave is damped and has a slightly lower frequency. The rate of damping depends on ζ and Fig. 5.18 also shows this response for various values of ζ.

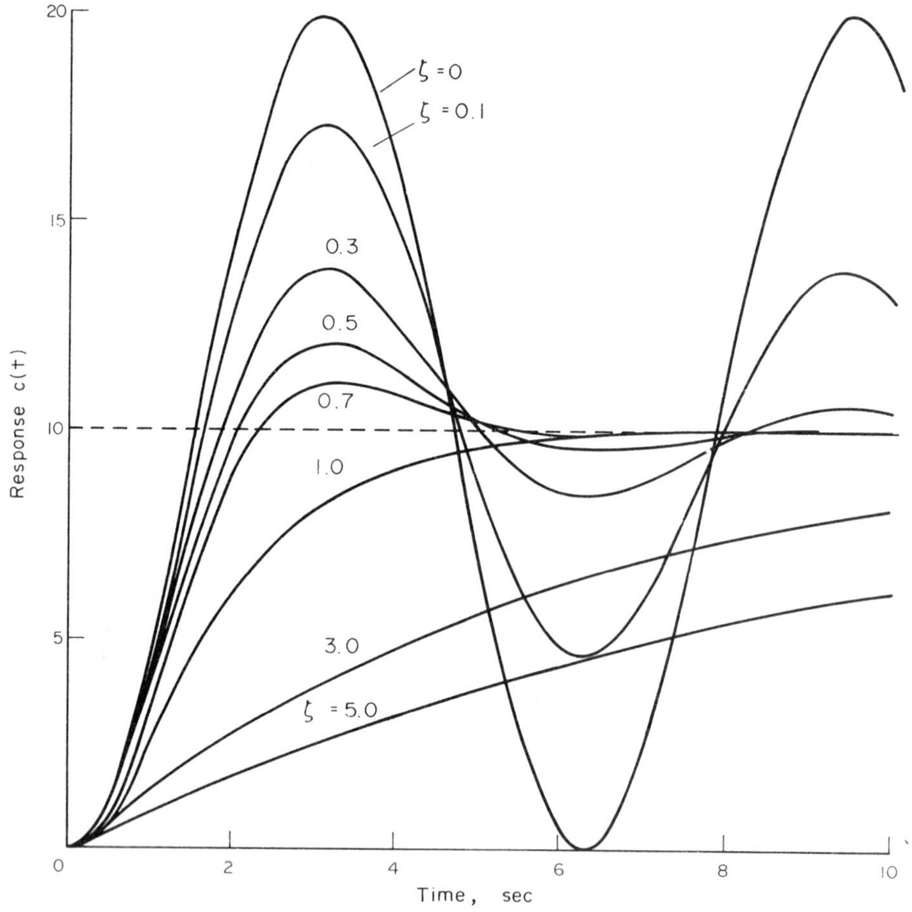

Fig. 5.18. Second-order step response with various values of ζ.

(c) $\zeta = 1$ - *critical damping*

This is the case in which the transform simplifies to that given in equation (5.37). This corresponds to entry no. 15 in Table 5.1, hence the time response is:

$$c(t) = 10(1 - (\omega_n t + 1)\cdot\exp(-\omega_n t))$$

which is shown in Fig. 5.18.

(d) $\zeta > 1$ - *overdamped case*

Here $\alpha = \sqrt{\zeta^2 - 1}$ is real and the response is non-oscillatory. Equation (5.36) becomes:

$$c(t) = 10 - 10e^{-\zeta\omega_n t}\left[\frac{\zeta}{\sqrt{\zeta^2 - 1}}\left(\frac{\exp(\omega_n\sqrt{\zeta^2 - 1})t - \exp(-\omega_n\sqrt{\zeta^2 - 1})t}{2}\right)\right.$$

$$\left. + \left(\frac{\exp(\omega_n\sqrt{\zeta^2 - 1})t + \exp(-\omega_n\sqrt{\zeta^2 - 1})t}{2}\right)\right]$$

$$= 10 - 10e^{-\zeta\omega_n t}\left[\frac{\zeta}{\sqrt{\zeta^2 - 1}}\sinh \omega_n\sqrt{\zeta^2 - 1}\,t + \cosh \omega_n\sqrt{\zeta^2 - 1}\,t\right]$$

since
$\sinh x = (e^x - e^{-x})/2$

and $\cosh x = (e^x - e^{-x})/2$.

This response is also shown in Fig. 5.18 for a number of values of ζ. It can be seen that as ζ gets larger, more and more damping produces an output response which is progressively more sluggish.

5.6.2 The Effects of Gain on Steady-state Response

The previous section dealt with the effect of controller gain on the shape of the dynamic response of the second-order system, via the effect of values of gain K. It will be recalled that equation (5.34) also included a multiplying term $K/(1 + K)$ and this has a scaling effect on the size of the response. Hence, as K increases, not only will the value of ζ decrease, to give a more oscillatory response, but also the final steady value reached will increase and approach the input step size. To fix ideas, we take the example of equation (5.33) with $T_1 = 10$ sec, $T_2 = 100$ sec. Figure 5.19 shows the manner in which the step response grows as K increases, and the damping reduces with increase in K. The undamped natural frequency ω_n and damping ratio ζ are related to T_1, T_2 and K by equations (5.34a) and (5.34b). For the underdamped response, the decay is fixed by the term $\exp(-\zeta\omega_n t)$ and the product $\zeta\omega_n$ is in fact constant, independent of the value of K.

5.6.3 Second-order Step Response Performance Specifications

The various step responses shown in Figs. 5.18 and 5.19 range from very sluggish but stable (with ζ greater than 2 or 3) to very rapid-acting but oscillatory, the oscillations dying away only slowly (ζ about 0.1 or 0.2). Neither of these extremes would be very acceptable as the step response of a practical controlled system. The best compromise appears to be a response such as that for ζ approximately 0.7, where the response is reasonably fast and the oscillations are not too large and die away fairly quickly. Because of this, this type of under-damped response is aimed for in practical second-order systems, and also in higher-order systems which can often be represented by equivalent second-order parameters.

Figure 5.20 illustrates some of the ways in which the performance of such a system

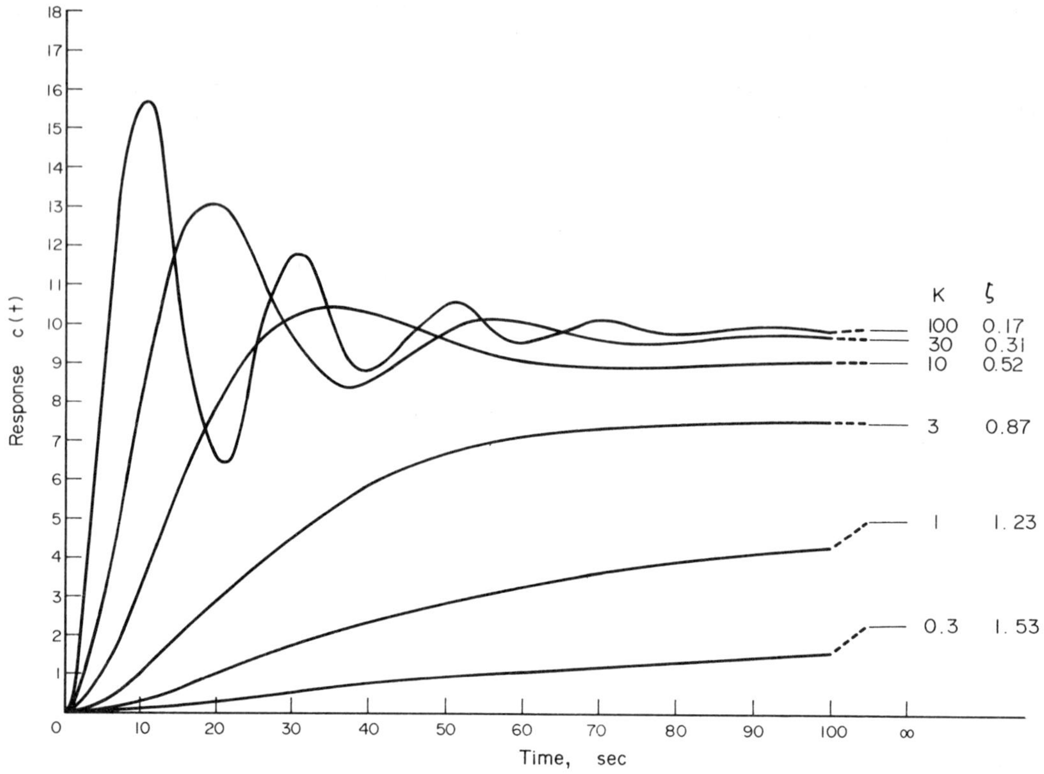

Fig. 5.19. Second-order step responses with gain effect.

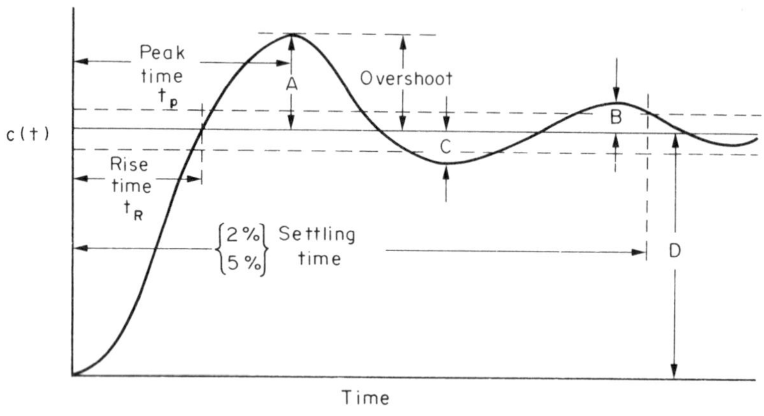

Fig. 5.20. General second-order damped step response.

may be specified. The specification used will depend on the application: with a servo-mechanism, for instance, it may be important to ensure that a certain speed of response is obtained; in such a case the rise time would be specified, whereas in a chemical process some maximum temperature may exist which must not be exceeded; the maximum overshoot would be specified in that case. In building-services control it is often desirable to limit the amount of oscillation of the control valve in order to reduce the wear, so a performance in terms of a required decay ratio or settling time could be stated. Most of the parameters of the performance can be related fairly simply to the damping ratio ζ and natural frequency ω_n of the second-order system, using the equation for the underdamped step response which was derived in Section 5.6.1(b). This equation is restated here for a final value of unity:

$$c(t) = 1 - \exp(-\zeta\omega_n t)\left[\frac{\zeta}{\sqrt{1-\zeta^2}}\sin\omega_d t + \cos\omega_d t\right] \qquad (5.38)$$

where $\omega_d = \omega_n\sqrt{1-\zeta^2}$ and $0 < \zeta < 1$.

Rise time t_R

This is the time taken for the final value to be first attained. It is also called the duplication time. From equation (5.38) we have, for $t = t_R$:

$$1 - \exp(-\zeta\omega_n t_R)\left[\frac{\zeta}{\sqrt{1-\zeta^2}}\sin\omega_d\, t_R + \cos\omega_d t_R\right] = 1.$$

Hence $\frac{\zeta}{\sqrt{1-\zeta^2}}\sin\omega_d t_R + \cos\omega_d t_R = 0$,

$\therefore \quad \tan\omega_d t_R = -\sqrt{1-\zeta^2}/\zeta$

$\tan\ \omega_d t_R$ is negative, hence $\pi/2 < \omega_d t_R < \pi$,

$\therefore \quad \tan(\pi - \omega_d t_R) = +\sqrt{1-\zeta^2}/\zeta$

$$t_R = \frac{\pi - \arctan\left(\sqrt{1-\zeta^2}/\zeta\right)}{\omega_n\sqrt{1-\zeta^2}}$$

The variation of the dimensionless product $\omega_n t_R$ with ζ is shown in Fig. 5.21.

Overshoot ratio and peak time t_p. The overshoot ratio is the ratio of the rise of the first peak above the final value to the final change: the ratio A/D in Fig. 5.20. The percentage overshoot ratio (overshoot ratio x 100%) is usually specified.

Differentiating equation (5.38) and setting $\dot{c}(t) = 0$ gives

$$\frac{\omega_n}{\sqrt{1-\zeta^2}}\exp(-\zeta\omega_n t)\sin\omega_d t = 0.$$

Hence $\sin\omega_d \cdot t = 0$,

$\therefore\ \dot{c}(t) = 0$ when $\omega_d \cdot t = 0, \pi, 2\pi, 3\pi$, etc.

The first value in this series, $\omega_d \cdot t_p = 0$, corresponds to $t = 0$, where the slope starts off at zero, so the first peak corresponds to $\omega_d \cdot t_p = \pi$.

The peak time, t_p, is thus given simply by

$$t_p = \pi/\omega_d$$

$$= \pi/\omega_n \sqrt{1 - \zeta^2}$$

and the value of the response at t_p is then

$$c(t_p) = 1 - \exp(-\zeta\omega_n t_p) \left[\frac{\zeta}{\sqrt{1 - \zeta^2}} \sin \pi + \cos \pi\right]$$

$$= 1 + \exp(-\zeta\omega_n t_p)$$

$$= 1 + \text{overshoot.}$$

Since $\sin \pi = 0$ and $\cos \pi = -1$.

Hence, overshoot ratio $= \exp\left[\frac{-\zeta\pi}{\sqrt{1 - \zeta^2}}\right]$.

The variations of $\omega_n \cdot t_p$ and overshoot ratio with damping ratio ζ are shown in Fig. 5.21.

Decay ratio. This is the ratio of successive positive (or successive negative) peaks, measured from the final value, corresponding to the ratio B/A in Fig. 5.20.

From the results obtained for peak time, we can write:

$$\text{Decay ratio} = B/A$$

$$= \frac{\exp(-3\zeta\pi/\sqrt{1 - \zeta^2})}{\exp(-\zeta\pi/\sqrt{1 - \zeta^2})}$$

$$= \exp(-2\zeta\pi/\sqrt{1 - \zeta^2}).$$

Decrement factor. This is an alternative way of specifying how quickly oscillations must die away. It is defined as the ratio of size of successive peaks measured from the final value, one a maximum, one a minimum, corresponding to the ratio A/C in Fig. 5.20. Note that the decrement factor is a number greater than unity, whereas the decay ratio is always less than unity. From the previous results, we have

$$\text{Decrement factor} = A/C$$

$$= \frac{\exp(-\zeta\pi/\sqrt{1 - \zeta^2})}{\exp(-2\zeta\pi/\sqrt{1 - \zeta^2})}$$

$$= \exp(+\zeta\pi / \sqrt{1 - \zeta^2}).$$

The decrement factor is occasionally stated in terms of the logarithmic decrement, which is defined as the natural logarithm of the decrement factor:

$$\text{Logarithmic decrement} = \ln (A/C)$$

$$= \zeta\pi/\sqrt{1 - \zeta^2}.$$

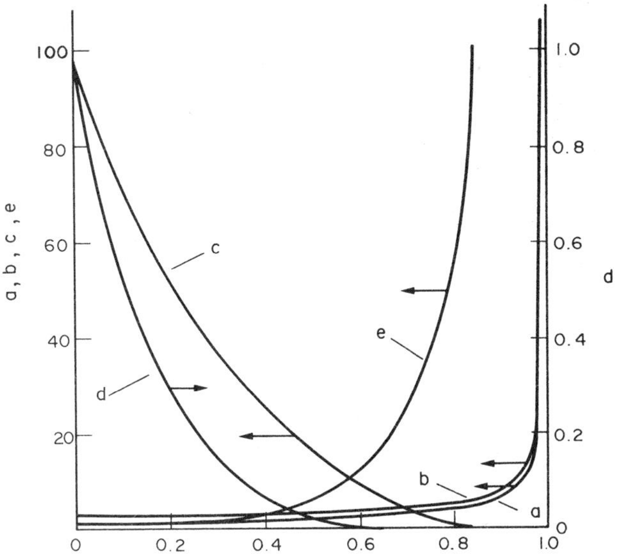

Fig. 5.21. Performance specifications for underdamped second-order step response. (a) $\omega_n \cdot t_R$, (b) $\omega_n \cdot t_p$, (c) % overshoot, (d) decay ratio, (e) decrement factor.

The decay ratio and decrement factor versus damping ratio ζ are also plotted in Fig. 5.21.

It can be seen from Fig. 5.21 that the rise time t_R and peak time t_p do not characterise the response very strongly: the curves (a) and (b) change only very slowly over most of the range of ζ up to a value of about 0.85, then increase extremely rapidly as ζ approaches 1.0. For small values of ζ, up to about 0.5, either the percentage overshoot or the decay ratio can be used since both are sensitive to small changes in ζ. Between about 0.5 and 0.8 the decrement factor changes rapidly with ζ.

Settling time t_s. The response can also be defined by specifying a time by which the oscillations must enter and stay within a given proportion of the final value. This time is known as the settling time and the proportion taken is either 5% or 2% of the final value. In Fig. 5.20 the peak of amplitude B is the last one for which the response will have exceeded the band shown by the dashed lines; subsequent oscillations will remain within that band. The value of settling time cannot be obtained analytically since it is a discontinuous function of ζ, but it can be given approximately as

$$t_s = 4/\zeta\omega_n \text{ for 2\% settling}$$

or

$$t_s = 3/\zeta\omega_n \text{ for 5\% settling.}$$

5.7 HIGHER-ORDER SYSTEMS AND STABILITY

Considering the first-order system of Section 5.5 and the second-order system of Section 5.6, it appears that the first-order system will always be stable whatever the value of gain; a second-order system response will also be unconditionally stable but small values of gain will give monotonic (non-oscillatory) response and as the gain increases then oscillations will set in and gradually die away. However large the gain, the oscillations will never build up in amplitude to give an unstable response. It is only for higher-order systems that the response can become unstable.

In order to examine the basic requirements for stability, we can consider a generalised closed-loop system such as that shown in Fig. 5.22, where all the forward path transfer functions have been combined into the single function $G(s)$ and all those in the feedback path into the function $H(s)$. We can write down the overall relationship between the input $r(s)$ and the output $c(s)$ as

$$c(s) = \left[\frac{G(s)}{1 + G(s)H(s)}\right] \cdot r(s). \tag{5.39}$$

The function $\left[\frac{G}{1 + GH}\right]$ is the Closed-Loop Transfer Function (CLTF).

The function GH is the Open-Loop Transfer Function (OLTF).

The function $1 + GH$ is called the characteristic function.

The equation $1 + GH = 0$ is called the characteristic equation.

We shall show that the nature of the response to any input is determined by the solutions of the characteristic equation.

In general the CLTF of a linear system can be written as a rational function of polynomials in s:

$$c(s) = \frac{a0 + a1s + a2s^2 + a3s^3 + \ldots + ans^n}{s^k(b0 + b1s + b2s^2 + b3s^3 + \ldots + bms^m)} \cdot r(s) \tag{5.40}$$

If we consider the pulse response of this system and let the n roots of the numerator be Z1, Z2, Z3, ..., Zn (because they are zeroes of the function), and the $k + m = q$ roots of the denominator be P1, P2, P3, ..., Pq (because they are poles of the function), we have

$$c(s) = \frac{K(s - Z1)(s - Z2)(s - Z3) \ldots (s - Zn) \cdot 1}{(s - P1)(s - P2)(s - P3) \ldots (s - Pq)}$$

$$= \frac{K \prod_{j=1}^{n} (s - Zj)}{\prod_{i=1}^{q} (s - Pi)} .$$

Note that the Pi, Zj may be zero, positive real or negative real or may occur in complex conjugate pairs.

To obtain the time response $c(t)$ to the impulse input, we expand $c(s)$ into partial fractions and invert:

$$c(s) = \frac{A1}{s - P1} + \frac{A2}{s - P2} + \ldots + \frac{Aq}{s - Pq}$$

$$= \sum_{i=1}^{q} \frac{Ai}{s - Pi}$$

The Ai may be determined by making use of Heaviside's theorem (Section 5.3.1), and since

$$L^{-1}\left(\frac{Ai}{s - Pi}\right) = Ai\ e^{Pit}$$

we can write

$$c(t) = A1\ e^{P_1 t} + A2\ e^{P_2 t} + A3\ e^{P_3 t} + \ldots + Aq\ e^{P_2 t}$$

$$= \sum_{i=1}^{q} Ai\ e^{Pit}$$

The magnitude of the response is determined by the values of the Ai, but the nature of the response is determined by the Pi. There are six possibilities:

(a) p zero, e.g. $Ae^{pt} = A$, giving a constant term, independent of time (see Fig. (5.23(a)).

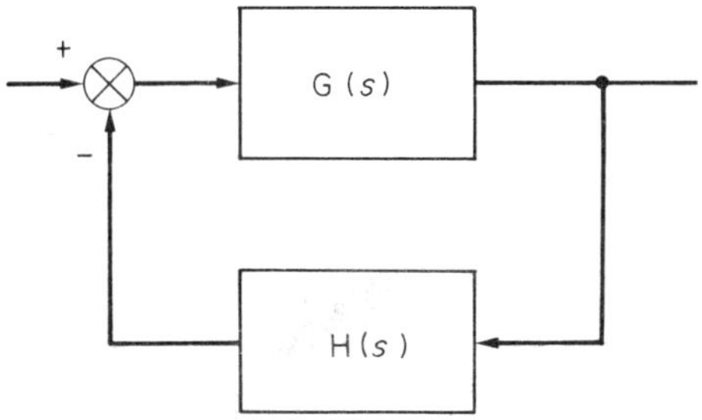

Fig. 5.22. General negative feedback system.

(b) p real and negative, e.g. $2e^{-3t}$. Here the term dies away exponentially without oscillations (see Fig. 5.23(b)). Stable.

(c) p complex, negative real part, e.g. 2 exp(-3 $\pm$ j2)t. Here the response is oscillatory with damped oscillations which die away with time (see Fig. 5.23(c)). Stable.

(d) p purely imaginary, zero real part, e.g. 2 exp($\pm$ j2t). This produces a sinusoidal response of constant amplitude (see Fig. 5.23(d)). Marginally stable.

(e) p complex, positive real part, e.g. 2 exp(3 $\pm$ j2)t. This has an oscillatory response the amplitude building up without limit (see Fig. 5.23(e)). Unstable.

(f) p real and positive, e.g. $2e^{3t}$. The response increases monotonically to

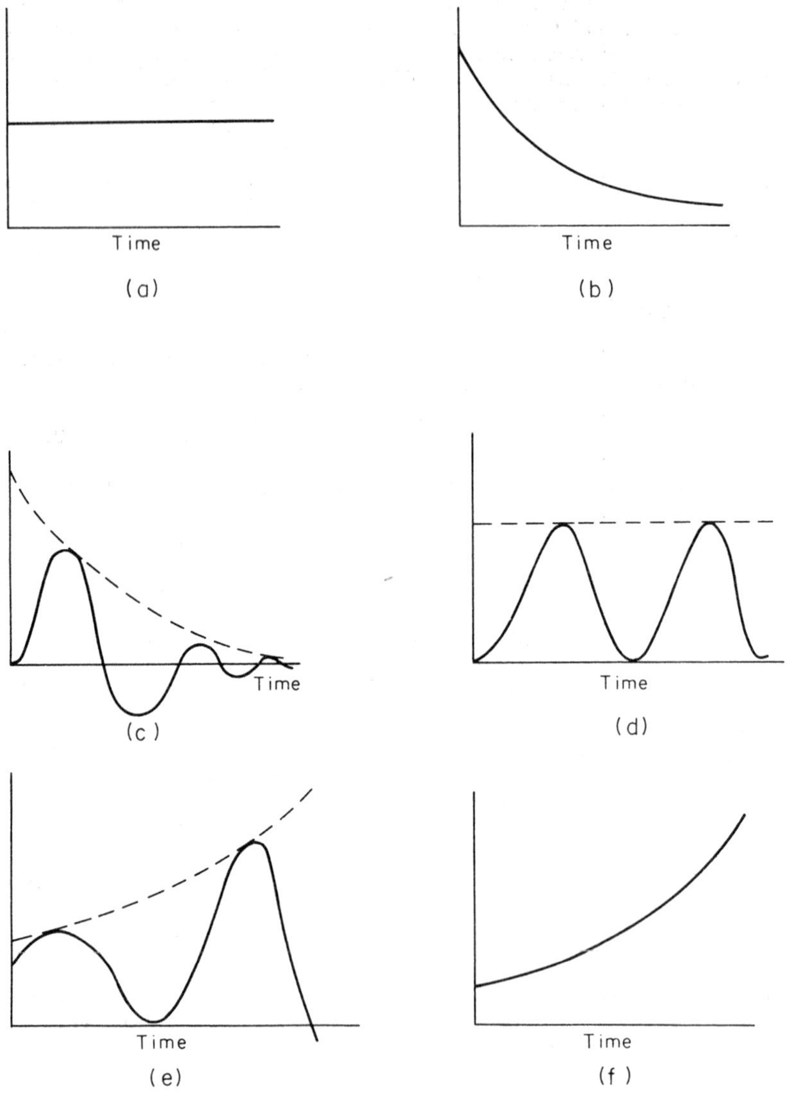

Fig. 5.23. The six possible types of dynamic response contributions.

infinity (see Fig. 5.23(f)). Unstable.

It can be concluded from the above analysis that the criterion of whether or not the original controlled system is stable is decided by testing whether any pole of the closed loop transfer function (or root of the characteristic function $1 + G(s)H(s)$) lies in the righthand half of the s-plane. Hence, in high-order systems we can test for stability of the time response indirectly by finding whether any root of the characteristic equation $1 + G(s)H(s) = 0$ has a positive real part. If there is at least one such root then the system is unstable. This is the basis of all stability criteria.

We shall briefly describe three of the stability criteria available for single-variable systems, one algebraic and two graphical. As a preliminary it is worth mentioning that for stability two simple conditions must be fulfilled: (a) all powers of s from s^n to s^0 must be present in the characteristic equation of an n-th-order system, and (b) all the coefficients must have the same sign. These are necessary but not sufficient conditions for stability.

For example, the equations $2s^5 + 6s^4 + 8s^2 + s + 1 = 0$ and $6s^4 + 10s^3 + 7s^2 - 2s + 3 = 0$ are both unstable because the above-mentioned conditions are not satisfied, but the equation $3s^3 + s^2 + 2s + 8 = 0$ is also unstable even though it satisfies both conditions.

5.8 ROUTH'S CRITERION

This procedure (the Hurwitz criterion is similar) takes the characteristic equation $1 + G(s)H(s) = 0$, expressed in polynomial form:

$$An\,s^n + An\text{-}1\,s^{n-1} + An\text{-}2\,s^{n-2} + \ldots + A2\,s^2 + A1\,s^1 + A0 = 0. \qquad (5.41)$$

From the coefficients An to Ao in that order we form the Routh Array:

$$
n+1 \text{ rows} \left\{ \begin{array}{c|ccccc}
s^n & An & An\text{-}2 & An\text{-}4 & An\text{-}6 & - \\
s^{n-1} & An\text{-}1 & An\text{-}3 & An\text{-}5 & An\text{-}7 & - \\
s^{n-2} & b1 & b2 & b3 & - & - \\
' & c1 & c2 & c3 & - & - \\
' & ' & ' & ' & & \\
' & ' & ' & ' & & \\
s^2 & & & & & \\
s^1 & j & & & & \\
s^0 & k & & & &
\end{array} \right.
$$

The first two rows are formed by writing the coefficients of the original characteristic equation in the "zig-zag" fashion indicated. The values of b1, b2, b3, etc., in the third row are evaluated from the formulae:

$$b1 = \frac{An\text{-}1\ An\text{-}2 - An\ An\text{-}3}{An\text{-}1},$$

$$b2 = \frac{An\text{-}1\ An\text{-}4 - An\ An\text{-}5}{An\text{-}1},$$

$$b3 = \frac{An\text{-}1\ An\text{-}6 - An\ An\text{-}7}{An\text{-}1}, \quad \text{etc.}$$

Successive values are formed by the same pattern of formula until the rest of the b's are all zero.

The fourth row is then formed using the values in the second and third rows:

$$c1 = \frac{b1\ An\text{-}3 - b2\ An\text{-}1}{b1},$$

$$c2 = \frac{b1\ An\text{-}5 - b3\ An\text{-}1}{b1},$$

$$c3 = \frac{b1\ An\text{-}7 - b4\ An\text{-}1}{b1}, \quad \text{etc.}$$

Further rows are similarly formed, continuing until each row is completed with zero elements and until n + 1 non-zero rows have been formed. Note that the s^1 and s^0 rows contain only one non-zero element each.

Routh's criterion is then stated as follows:

The number of roots of the characteristic equation with positive real parts is equal to the number of changes of sign in the first column of the array.

Hence the system is stable if the elements of the first column:

An
An-1
b1
c1
'
'
'
'
j
k

are all positive or all negative.

EXAMPLE 5.4. Applying the Routh criterion to the characteristic equation $2s^4 + s^3 + 12s^2 + 8s + 2 = 0$ we have:

s^4	2	12	2
s^3	1	8	
s^2	-4	2	
s^1	8.5		
s^0	2		

There are two sign-changes in the first column (from positive to negative, then from negative back to positive), so the equation has two roots in the right-hand half of the s-plane and is thus unstable.

The roots are in fact

$$-0.34 \pm j0.23$$

and

$$+0.089 \pm j2.44$$

which confirms the above result.

As well as simply indicating whether a given system is stable or not, this criterion can be used to find the limits on values of the coefficients for stability.

EXAMPLE 5.5. For the characteristic equation

$$s^3 + 10s^2 + 24s + K = 0$$

constructing the Routh array gives

s^3	1	24
s^2	10	K
s^1	$\frac{240-K}{10}$	
s^0	K	

Hence, the system represented by this equation will be stable for $0 < K < 240$ since it is only for K within this range that the elements of the first row are all positive.

If a common positive factor exists between the elements of a row, they may be divided by this factor without changing the signs of the first column. This can simplify the evaluation of the subsequent elements.

It sometimes happens that either a single zero element occurs in the first column, or all the elements of a row become zero. In these cases the evaluation of further elements cannot proceed directly, but two simple techniques can be used to complete the array.

(i) With a zero in the first column, the zero can be replaced by a small positive number, δ, and the array completed. The change in signs is assessed by observing the signs of the limits as the magnitude of δ is allowed to approach zero. Alternatively, the order of the coefficients may be reversed to give the same result.

EXAMPLE 5.6. For the characteristic equation

$$s^5 + 2s^4 + 2s^3 + 4s^2 + s + 1 = 0$$

constructing the array gives:

s^5	1	2	1
s^4	2	4	1
s^3	ϕ δ	$\frac{1}{2}$	
s^2	$(4\delta - 1)/\delta$	1	
s^1	A	0	
s^0	1	0	

where $A = (4\delta - 1 - 2\delta^2)/(4\delta - 1)$

As δ approaches zero $(4\delta - 1)/\delta$ approaches $-1/\delta$ which is negative and A approaches +1, so the system is unstable.

Reversing the coefficients gives the array:

s^5	1	4	2
s^4	1	2	1
s^3	-2	1	0
s^2	2.5	1	
s^1	1.8	0	
s^0	1	0	

which confirms the previous result.

(ii) With a vanishing row, in which all the elements are zero, an auxiliary equation is formed from the elements of the preceding row. The auxiliary equation is differentiated and the vanishing row is replaced by the coefficients of the differentiated equation. The construction of the array then continues as usual.

EXAMPLE 5.7. For the characteristic equation

$$s^4 + 2s^3 + 11s^2 + 18s + 18 = 0$$

we have the array:

s^4	1	11	18
s^3	2	18	0
s^2	2	18	
s^1	0	0	
(s^1	4	0)	
s^0	18	0	

when the s^1 row vanishes, the auxiliary equation is formed from the preceding row, the powers of s decreasing by two in successive terms:

$$2s^2 + 18 = 0$$

Differentiating this equation gives

$$4s + 0 = 0$$

whose coefficients are used to form a new s^1 row to complete the array. There are no sign changes in the first column, so a stable system is predicted.

The occurrence of an all-zero row indicates the presence of roots which are the negatives of each other, lying on a circle or circles centred on the origin. The roots of the auxiliary equation are also roots of the original equation. In this case the roots of the auxiliary equation are $\pm j3$, which lie on the imaginary axis. The equation is thus on the borderline between stability and instability, a fact which does not emerge from the Routh-Hurwitz analysis. For design purposes it is desirable to have some method which indicates how near a system is to instability and permits some assessment of the performance of a stable system. This is provided by the Root Locus method and the Nyquist criterion.

5.9 ROOTS OF POLYNOMIAL EQUATIONS

The Routh-Hurwitz procedure permits assessment of the stability or otherwise of a given system. It does not indicate how close to instability a stable system is or how close to stability an unstable system is. Similarly it gives no indication of whether the performance of a stable system is likely to be satisfactory. In order to arrive at such an indication it would be necessary to observe the movement of the characteristic equation roots as some parameter such as controller gain is varied, and this is possible using the Root Locus method as well as the Nyquist criterion. As a preliminary it will be useful to introduce some simple numerical methods for finding the roots of polynomial equations.

Newton-Raphson method. This can give the values of real roots only. Referring to Fig. 5.24, the true root is as shown where f(x) crosses the x-axis. An initial

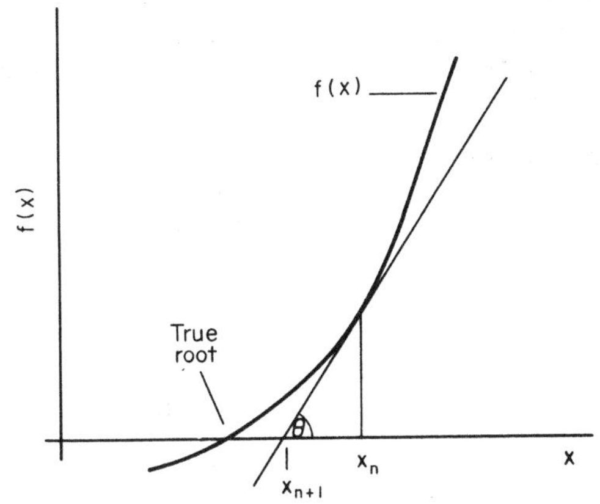

Fig. 5.24. Newton-Raphson method for real roots.

approximation to the root is x_n. At this value of x the tangent to the curve meets the x-axis at a value x_{n+1} and an angle θ. We can write for the rate of change of f(x) at $x = x_n$:

$$\tan\theta = f'(x_n) = \frac{f(x_n)}{x_n - x_{n+1}} = \frac{df(x)}{dx},$$

rearranging gives $x_{n+1} = x_n - f(x_n)/f'(x_n)$

so that x_{n+1} is an improved estimate of the root.

EXAMPLE 5.8. Find the real root of $f(x) = x^3 + 2x^2 + 4x + 6$.

Differentiating, $f'(x) = 3x^2 + 4x + 4$.

Taking an initial trial value $x_0 = 0$, we have $f(0) = 6$ and $f'(0) = 4$. Hence the next trial value x_1 is given by

$$x_1 = 0 - 6/4 = -1.5.$$

For the next stage, $f(-1.5) = 1.0$ and $f'(-1.5) = 4.75$, so $x_2 = -1.5 - 1/4.75 = -1.71$.

Continuing, $f(-1.71) = 0.01$, $f'(-1.71) = 5.93$,

$$x_3 = -1.71 - 0.01/5.93 = -1.71$$

so there is a real root at $x = -1.71$ and $(x + 1.71)$ is a factor.

Dividing $f(x)$ by this factor gives the remaining factor, a quadratic.

Lin's method. This method extracts quadratic factors and hence permits the evaluation of complex roots.

Taking the general polynomial:

$$An x^n + An\text{-}1 x^{n-1} + An\text{-}2 x^{n-2} + \ldots + A2x^2 + A1x + A0 = 0$$

a first trial factor is formed from the last three terms of the polynomial:

$$x^2 + (A1/A2)x + (A0/A2)$$

Denoting this factor by $x^2 + a1x + a0$ we carry out the long division:

$$\begin{array}{r|l} & Anx^{n-2} + \ldots + B2 \\ \hline x^2 + a1x + a0 & Anx^n + An\text{-}1x^{n-1} + \ldots + A2x^2 + A1x + A0 \\ & \underline{Anx^n + \ldots} \\ & \Rightarrow B2x^2 + B1x + A0 \Leftarrow \\ & \underline{B2x^2 + C1x + C0} \\ & \qquad D1x + D0 \end{array}$$

If the remainder $D1x + D0$ is too large then the next trial divisor is formed from the arrowed line: $x^2 + (B1/B2)x + (A0/B2)$. The process is repeated until the remainder is acceptably small.

EXAMPLE 5.9. Find the roots of the equation

$$x^4 + 4.5x^3 + 12.5x^2 + 17.5x + 12.5 = 0.$$

The first trial divisor is

$$x^2 + (17.5/12.5)x + (12.5/12.5) = x^2 + 1.4x + 1.$$

Dividing out:

$$\begin{array}{r|l} & x^2 + 3.1x + 7.16 \\ \hline x^2 + 1.4x + 1 & x^4 + 4.5x^3 + 12.5x^2 + 17.5x + 12.5 \\ & \qquad 7.16x^2 + 14.4x + 12.5 \\ & \qquad \underline{7.16x^2 + 10.0x + 7.16} \\ & \qquad\qquad 4.4x + 5.34 \end{array}$$

The second trial divisor is

$$x^2 + (14.4/7.16)x + (12.5/7.16) = x^2 + 2.01x + 1.75.$$

The process can be summarised in the following table:

Trial no.	Divisor	Quotient	Remainder
1	$x^2 + 1.4x + 1$	$x^2 + 3.1x + 7.2$	$4.4x + 5.3$
2	$x^2 + 2.01x + 1.75$	$x^2 + 2.5x + 2.5$	$1.6x + 2.2$
3	$x^2 + 2.3x + 2.2$	$x^2 + 2.2x + 5.3$	$0.6x + 1.0$
4	$x^2 + 2.4x + 2.4$	$x^2 + 2.1x + 5.1$	$0.3x + 0.4$
5	$x^2 + 2.5x + 2.5$	$x^2 + 2x + 5$	

Factorising the quadratic terms gives the roots: $-1.25 \pm j0.97$ and $-1 \pm j2$.

5.9.1 Root Locus Plots

For a polynomial equation with real coefficients, the roots will always be either real or occur in complex conjugate pairs. The root locus plot is the plot in the complex plane of the paths (loci) followed by roots as a parameter of the equation varies, usually over the range zero to positive infinity.

In general, for a negative feedback control system with an overall forward-path transfer function $G(s)$ and an overall feedback path transfer function $H(s)$, as in Fig. 5.22, the characteristic equation is

$$G(s)\cdot H(s) + 1 = 0,$$

hence

$$G(s)\cdot H(s) = -1$$

giving the two relations:

$$\text{Magnitude condition: } |G(s)\cdot H(s)| = 1$$

and,

$$\text{Angle condition: } \angle G(s)\cdot H(s) = \pm 180^\circ.$$

if $G(s)\cdot H(s)$ is regarded as a complex function whose value is determined by the value of the complex variable s.

From these two conditions a number of simple construction rules can be derived for sketching the loci of roots of a given equation, but before presenting these, a simple example will be shown, using only the quadratic formula.

EXAMPLE 5.10. For the system whose block diagram is shown in Fig. 5.25(a), we have

$$\text{characteristic equation: } s^2 + 2s + K = 0.$$

Since this is quadratic, it is straightforward to determine the roots by formula:

$$s_{1,2} = -1 \pm \sqrt{1 - K}.$$

Substituting values of K permits the evaluation of the roots directly.

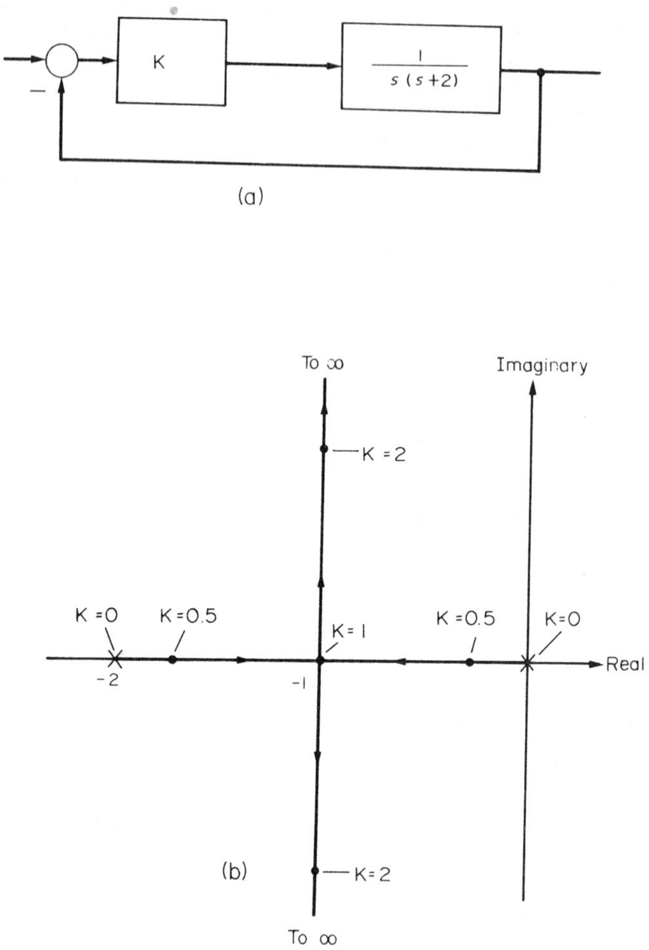

Fig. 5.25. A second-order system (a) and its root-locus plot (b).

$$K = 0 \quad : s = 0 \text{ or } -2,$$
$$K = 0.5 : s = -0.293 \text{ or } -1.707,$$
$$K = 1.0 : s = -1 \text{ repeated},$$
$$K = 2.0 : s = -1 \pm j,$$
$$K = 10.0 + s = -1 \pm j3.$$

These results are plotted in Fig. 5.25(b). It is evident that no value of K, however large, will drive the roots into the right-hand half of the complex plane. For $K > 1$ the roots are complex conjugate and the response to say a step input will be oscillatory with a damping coefficient between zero and one. In a later section it is shown how the actual values of ζ and ω_n may be obtained for a specific case. In this simple example the roots could easily be determined explicitly. For higher-order systems it becomes necessary to use the construction rules.

Rules for the construction of Root Locus plots

1. The number of branches of the locus is equal to the number of poles of the open-loop transfer function (OLTF), which is equal to the order of the characteristic equation.

2. The root locus exists on those portions of the real axis for which the number of poles and zeros to the right is odd.

3. Start points: the locus start points ($K = 0$) are at the OLTF poles. End points: the locus end points ($K = \infty$) are at the OLTF zeros or at infinity.

4. Asymptote angles: the angles of the asymptotes of the root locus that end at infinity are given by:

$$\frac{(1 + 2n)\ 180^\circ}{\text{Rank of OLTF}} \quad \text{for K positive,}$$

or by

$$\frac{n \cdot 360^\circ}{\text{Rank of OLTF}} \quad \text{for K negative.}$$

where $n = 0, \pm 1, \pm 2, \pm 3$, etc. The rank of the OLTF is the difference between the highest power of s in the denominator and the highest power of s in the numerator.

5. Real-axis intercept of asymptotes:

$$\text{Intercept} = \frac{\Sigma(\text{poles of OLTF}) - \Sigma(\text{zeros of OLTF})}{\text{Rank of OLTF}}$$

6. Breakaway points on the real axis: these may be determined by three methods.

   (i) setting $dK/ds = 0$ and solving,

   (ii) plotting K as a function of s and finding a maximum, or

   (iii) taking a trial point and applying the angle condition.

7. Angle of departure from a complex pole = $180^\circ - \Sigma$ (angles from other poles) $+ \Sigma$ (angles from other zeros).

8. Angle of approach to a complex zero = Σ (angles from other poles) $- \Sigma$ (angles from other zeros) $- 180^\circ$.

9. For a characteristic equation of the form

$$s^n + An\text{-}1\, s^{n-1} + An\text{-}2\, s^{n-2} + \ldots + A_1 s + A0 = 0$$

then Σ (real parts of roots) = $-An\text{-}1$

and: (product of roots) = $A0(-1)^n$.

10. Application of the angle condition:

For any trial point A, if

Σ (angles from poles to A) $- \Sigma$ (angles from zeros to A) $= (1+2n)180^\circ$ ($n = 0, \pm 1, \pm 2$, etc.),

then A is a point on the root locus.

To illustrate the use of these rules they will be applied to obtaining the root locus plots for some examples.

EXAMPLE 5.11. Draw the root locus plot for the system with OLTF: $K/s(s + 4)(s + 6)$ for positive values of K and hence find the minimum value of K to give an oscillatory response.

This open-loop transfer function has three poles: 0, -4, -6 and no zeros, so we expect three branches for the root locus plot (Rule 1).

From Rule 2 we expect the locus to exist on the real axis between the origin and -4 and between -6 and $-\infty$.

Rule 3: since there are no OLTF zeros we expect the locus branches to end at infinity.

Rule 4: In this case: Rank of OLTF = 3.

giving asymptote angles of $(1 + 2n)\cdot 60^{o}$, which generates a repeated cycle of angles at $\pm 60^{o}$ and $\pm 180^{o}$.

Rule 5: Intercept $= \dfrac{0 - 4 - 6}{3} = -3.33.$

The asymptote positions are fixed by the combination of this rule with the previous one. They are shown as dashed lines in Fig. 5.26.

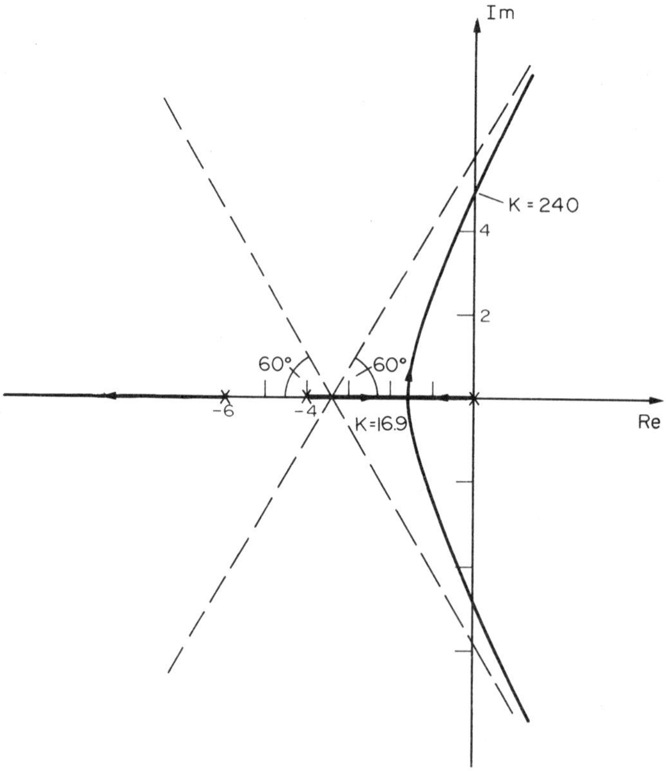

Fig. 5.26. Root locus plot for a third-order system (Example 5.11).

Rule 6: Since this is a third-order system it is straightforward to use method 6(i). The characteristic equation is

$$K/s(s + 4)(s + 6) + 1 = 0.$$

Rearranging and multiplying out:

$$K = -s^3 - 10s^2 - 24s, \qquad (5.42)$$

$$dK/ds = - 3s^2 - 20s - 24.$$

Setting dK/ds equal to zero and solving the resulting quadratic gives

$$s = -1.57 \text{ or } -5.10.$$

The first value is the correct breakaway point for positive values of K. The second value exists only on the locus for negative K and is a minimum, actually a break-in point. Substituting $s = -1.57$ in the characteristic equation gives $K = 16.9$ as the minimum value of gain for an oscillatory type of response. For values of K greater than this there will be a complex conjugate pair or roots.

Rules 7 to 10 do not apply to this case, and we already have enough information to sketch the locus, which is shown in Fig. 5.26.

The characteristic equation for this example is the same as that in Example 5.5, from which the limiting value of K for stable operation is seen to be 240. Setting K equal to 240 gives a zero row in the Routh array. The auxiliary equation obtained from the s^2 row is then $10s^2 + 240 = 0$ which has roots at $s = \pm j4.9$. These are the crossing points of the imaginary axis.

Design by equivalent second-order parameters. It often happens in a high-order system that one pair of roots lie closer to the origin than any of the others. Their contribution to the transient response will decay much more slowly than that of the other roots. The response of the system, which may in fact be of third or higher order, will therefore approximate to that of the second-order system represented by that pair of roots. They are called dominant roots and the transient performance of the system may be specified in terms of the equivalent second-order system.

Referring to the standard form of second-order transfer function, such as that in Equation (5.34), we can write the characteristic equation of such a system:

$$s^2 + 2\zeta\omega_n s + \omega_n^2 = 0.$$

The roots are:

$$s = \frac{-2\zeta\omega_n \pm \sqrt{4\zeta^2\omega_n^2 - 4\omega_n^2}}{2}$$

$$= - \zeta\omega_n \pm \omega_n \sqrt{\zeta^2 - 1}.$$

For $\zeta > 1$ (overdamped case) this gives a pair of roots on the negative real axis. The case where $\zeta < 1$ is usually of more interest, with a complex conjugate pair of roots.

$$s = -\zeta\omega_n \pm j\omega_n \sqrt{1 - \zeta^2}$$

$$= -\zeta\omega_n \pm j\omega_d.$$

Converting this to polar form, represented as points a distance r from the origin, at an angle θ from the real axis, gives

$$r = \sqrt{\zeta^2\omega_n^2 + \omega_n^2(1 - \zeta^2)} = \omega_n$$

$$\text{and } \cos\theta = \zeta\omega_n/\omega_n = \zeta,$$

$$\theta = \cos^{-1}\zeta \quad \text{(see Fig. 5.27).}$$

From these results we can identify lines in the s-plane which represent a constant value of some property of the response, such as ζ or ω_n. Lines of constant ζ will be lines of constant $\theta = \cos^{-1}\zeta$ - radial lines radiating from the origin. Lines of constant ω_d will be parallel to the real axis. Lines of constant $\zeta\omega_n$ (decay rate) will be parallel to the imaginary axis, and lines of constant ω_n will be circles centred on the origin. Intersection of lines indicates a combination of two or more such properties. These lines are indicated in Fig. 5.28. It should be emphasised that they strictly apply only to the response of a second-order system.

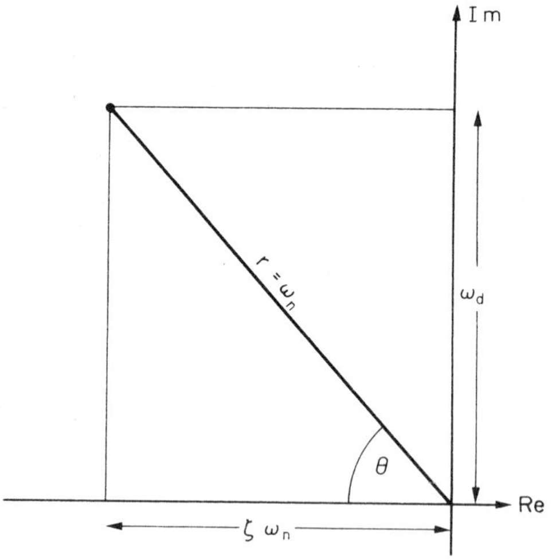

Fig. 5.27. Second-order root position.

EXAMPLE 5.12. Determine the value of gain K to give an equivalent damping ratio ζ of 0.6 in the system of Example 5.11. Find the equivalent values of ω_n and ω_d.

For a damping ratio of 0.6 we have a radial line making an angle $\cos^{-1} 0.6 = 53^o$ with the real axis. Superimposing this on the root locus plot of Fig. 5.26 gives the intersection with the root locus, as shown in Fig. 5.29. The intersection at point A has the coordinates $s = -1.3 + j1.7$. Substituting this value of s into equation (5.42) gives the (real) value of K as 34. The equivalent value of ω_n is the radial distance of this intersection point from the origin, which is

$$\sqrt{1.3^2 + 1.7^2} = 2.1 \text{ rad/sec.},$$

and the equivalent value of ω_d is the imaginary part of the intersection point A, which is 1.7 rad/sec. Hence we would expect the transient oscillations of the system with this value of K to decay according to the exponential $\exp(-\zeta\omega_n t)$ =

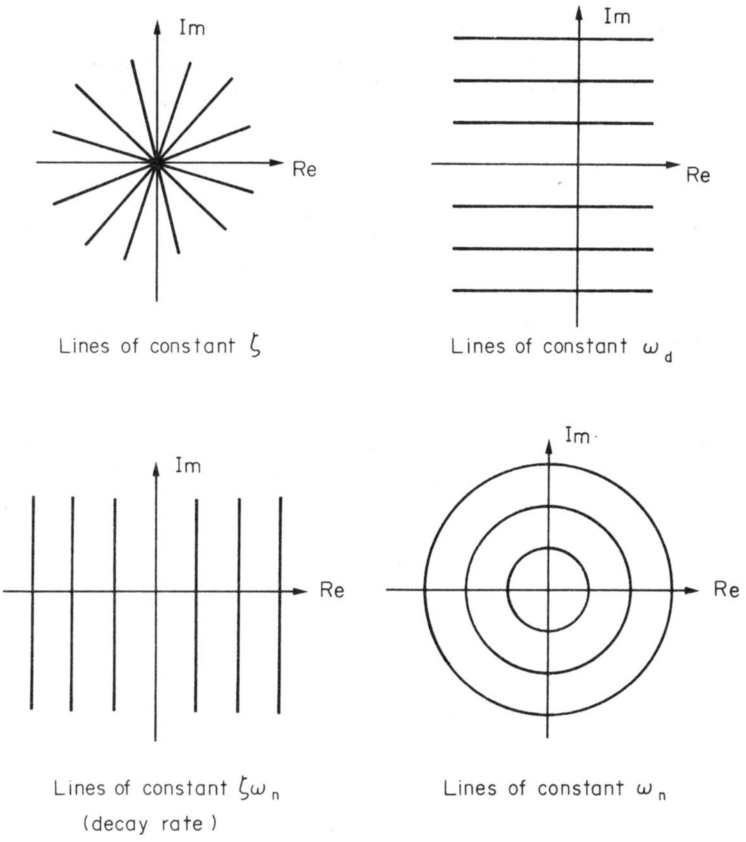

Fig. 5.28. Geometry determining second-order root positions.

exp(-1.26t).

5.9.2 Distance-Velocity Lag; Padé Approximation

Consider a hot-water boiler with a flow-water temperature $\theta(t)$ which varies with time (Fig. 5.30). The water temperature is sensed by three sensors, one within the boiler and two at points along the flow pipe. The three signals from the sensors may be denoted by $\theta_1(t)$, $\theta_2(t)$ and $\theta_3(t)$. The pattern of the sensor outputs will appear as in Fig. 5.31, the time delay between successive curves being given by the ratio: (distance between sensors)/(velocity of water in the pipe). The shapes of the three curves are identical; they are shifted in time. If the time delay between $\theta_1(t)$ and $\theta_2(t)$ is T_1, so that

$$\theta_2(t) = \theta_1(t - T_1),$$

then we may apply the shift theorem (Section 5.3.3) to the transforms of these functions:

$$L\theta_1(t) = \theta_1(s)$$

$$\therefore\ L\theta_2(t) = e^{-sT_1}\ \theta_1(s) = \theta_2(s).$$

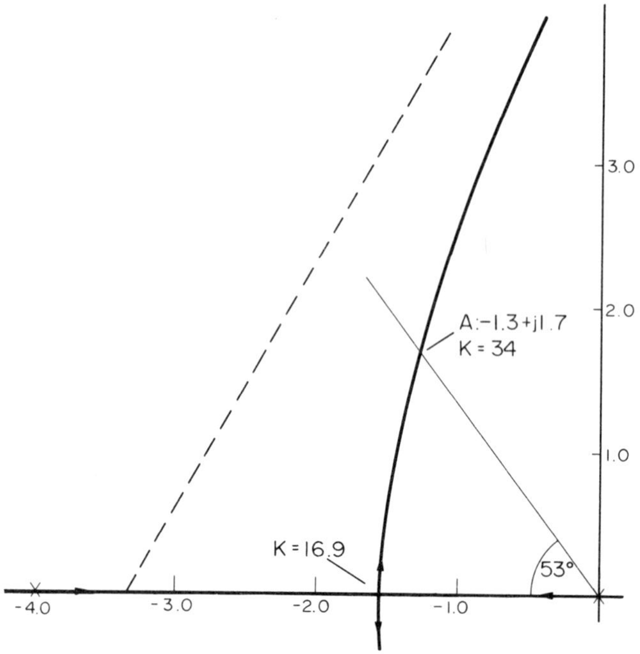

Fig. 5.29. Root locus for Example 5.12.

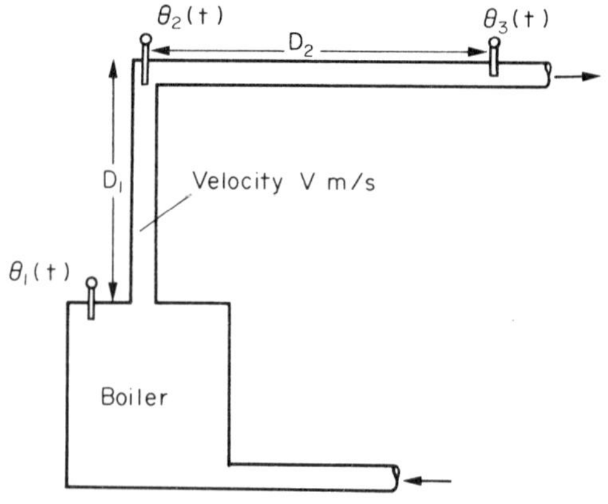

Fig. 5.30. Distance-velocity lag.

Similarly $$\theta_3(s) = e^{-sT_2}\,\theta_2(s) = e^{-sT_3}\,\theta_1(s)$$

where $$T_3 = T_2 + T_1.$$

This arrangement occurs very frequently in building services systems, giving rise to transfer functions incorporating exponentials of this type.

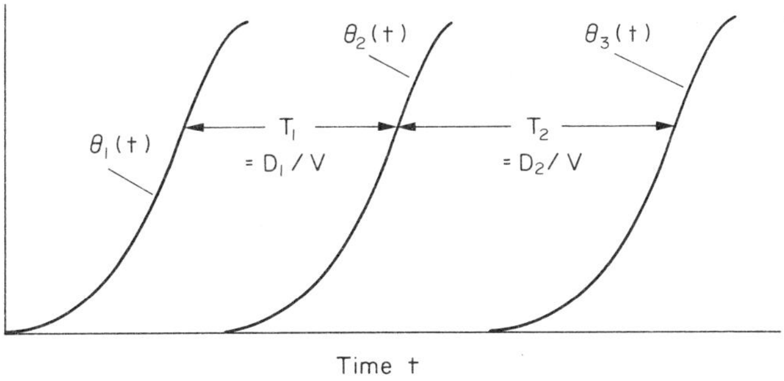

Fig. 5.31. Signals from sensors with distance-velocity lags.

Since the exponential function can be expressed as an infinite series:

$$e^x = 1 + x/1! + x^2/2! + \dots = \sum_{n=0}^{\infty} x^n/n!$$

the transfer function for the distance-velocity lag can be written in the form of a ratio of two series:

$$e^{-sT} = \frac{\exp(-sT/2)}{\exp(+sT/2)}$$

$$= \frac{1 - sT/2 + s^2T^2/8 - s^3T^3/48 + \dots}{1 + sT/2 + s^2T^2/8 + s^3T^3/48 + \dots}$$

and the function can be approximated by truncating the numerator and denominator series after a chosen number of terms. This is the Padé approximation. The first-order approximation includes only the first two terms in each series:

$$e^{-sT} \approx (1 - sT/2)/(1 + sT/2) = (2 - sT)/(2 + sT).$$

EXAMPLE 5.13. Using the Routh-Hurwitz criterion, investigate the limiting values of gain K, for the two open-loop transfer functions.

$$K/(1 + s)^3 \text{ and } K\, e^{-s}/(1 + s)^3.$$

For the first case the characteristic equation is

$$K/(1 + s)^3 + 1 = 0.$$

Expanding gives:

$$s^3 + 3s^2 + 3s + 1 + K = 0.$$

which has a Routh array as follows:

s^3	1	3
s^2	3	1 + K
s^1	(8-K)/3	0
s^0	1 + K	

From the s^1 row we see that for stability K must be less than 8.

For the second case the characteristic equation is

$$s^3 + 3s^2 + 3s + 1 + K\,e^{-s} = 0.$$

Replacing the exponential function by the first-order Padé approximation gives

$$s^3 + 3s^2 + 3s + 1 + K(2 - s)/(2 + s) = 0$$

which corresponds to

$$s^4 + 5s^3 + 9s^2 + (7 - K)s + 2(1 + K) = 0$$

which has the Routh array as follows:

s^4	1	9	2 + 2K
s^3	5	7 - K	0
s^2	(38+K)/5	2 + 2K	
s^1	A	0	
s^0	2 + 2K		

where A = (2.58 - K)(83.6 + K)/(38 + K). For stability, K must not exceed 2.58.

It can be shown by using the Nyquist criterion (see Section 5.10.2) that the true limit on K for stability in this second case is 2.49. The first-order Padé approximation gives an acceptable result in this case. It shows that the presence of the distance-velocity lag severely limits the likelihood of stability.

Because the exponential function can be regarded as equivalent to a polynomial of infinite degree, many of the rules given previously for construction of the Root-locus plot cannot be applied, and it is necessary to construct the plot by having recourse to the Angle Condition directly. This is very tedious as a manual procedure since the plots must be constructed point by point, but it can be handled fairly conveniently by digital computer programs.

5.10 FREQUENCY RESPONSE PLOTS

As an alternative to the Routh-Hurwitz and Root-Locus methods of analysis, it is also possible to predict a control systems behaviour by examining its sinusoidal frequency response. This uses the Nyquist Criterion. As a preliminary, it will be convenient to consider how the frequency response of a system may be obtained from its transfer function directly.

For a system with a Laplace transfer function F(s), the output response can be obtained by the general method outlined in Section 5.3, so that

$$y(s) = F(s)\cdot x(s)$$

where x and y are the input and output functions respectively. For a sinusoidal input function of unit magnitude:

$$x(t) = \sin \omega t$$

and from Table 5.1 we have

$$x(s) = \omega/(s^2 + \omega^2)$$

so that the output transform is obtained from:

$$y(s) = F(s)\cdot\omega/(s^2 + \omega^2)$$

and the time response y(t) may be obtained by inverse transformation, using one of the methods described in Section 5.3.1. This must then be repeated for various values of frequency over a suitable range.

In fact this procedure is unnecessary, because a sinusoidal signal happens to be the only one which passes unaltered in form through a linear system: a sine wave input, in the steady state, produces a sine wave output of generally a different amplitude and phase from that of the input. The frequency response of a linear system is therefore completely specified by the manner in which the amplitude ratio and the phase difference between input and output vary with applied frequency ω. With reference to Fig. 5.32, the response at a particular frequency, ω_1, is given by

$$\text{Magnitude ratio} = M(\omega_1) = Y(\omega_1)/X(\omega_1)$$

$$\text{Phase difference} = \phi\ (\omega_1)$$

It will now be shown that the magnitude ratio and phase difference for a specific frequency ω_1 can be found by substituting $j\omega_1$ for s wherever it appears in the transfer function $F(s)$.

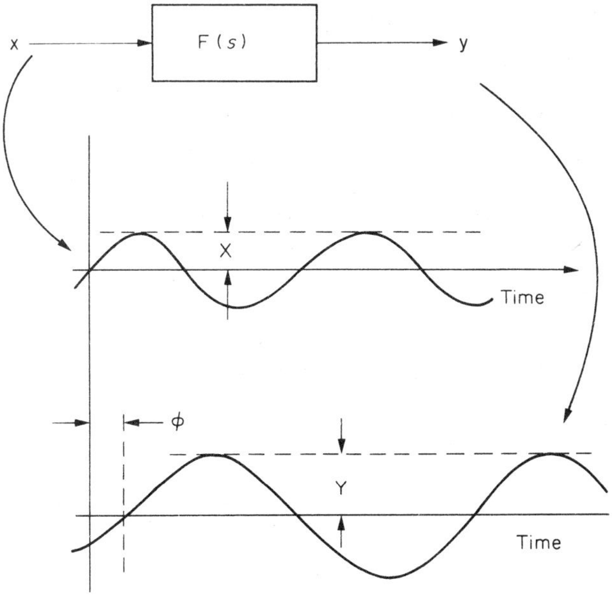

Fig. 5.32. Sinusoidal input and output of a linear system.

Then $$M(\omega_1) = |F(j\omega_1)|$$

and $$\phi(\omega_1) = \underline{/F(j\omega_1)}.$$

Taking the general rational transfer function $F(s)$ to be in factored form:

$$F(s) = \frac{K(s - a_1)(s - a_2)(s - a_3) \ldots (s - a_m)}{(s - b_1)(s - b_2)(s - b_3) \ldots (s - b_n)}$$

the sinusoidal response $y(s)$ is given by:

$$y(s) = F(s)\ X\omega/(s^2 + \omega^2)$$

$$= \frac{K(s - a_1)(s - a_2) \ldots (s - a_m)\omega}{(s - b_1)(s - b_2) \ldots (s - b_n)(s + j\omega)(s - j\omega)}$$

where the input amplitude, X, has been taken as unity and the quadratic term has been factored out.

If $n > m$, which is the case for all real physical systems, we can expand into partial fractions:

$$y(s) = \frac{A_1}{s - b_1} + \frac{A_2}{s - b_3} + \frac{A_3}{s - b_3} + \ldots + \frac{A_n}{s - b_n} + \frac{A'}{s - j\omega} + \frac{A''}{s + j\omega}.$$

The inversion of each of these terms is straightforward, giving the time response of form:

$$y(t) = A_1e^{b_1t} + A_2e^{b_2t} + \ldots + A_ne^{b_nt} + A'e^{j\omega t} + A''e^{-j\omega t}$$

If the system is stable then all the values of b (the poles of $F(s)$) are negative and real. The first n terms above will die away with time and eventually in the steady state the response will be

$$y(t) = A'e^{j\omega t} + A''e^{-j\omega t} \qquad (t >> 1)$$

The two coefficients A' and A" can be found from Heaviside's expansion:

$$A' = \left[\frac{(s - j\omega)F(s)\omega}{s^2 + \omega^2}\right]_{s = +j\omega}$$

$$= \left[\frac{F(s)\omega}{(s + j\omega)}\right]_{s = +j\omega} = \frac{F(j\omega)}{2j},$$

similarly $$A'' = \frac{-F(-j\omega)}{2j}.$$

Hence $$y(t) = \frac{F(j\omega)e^{j\omega t} - F(-j\omega)e^{-j\omega t}}{2j}.$$

Now let $$F(j\omega) = \alpha + j\beta = |F(j\omega)|\ e^{j\phi}$$

and $$|F(-j\omega)| = \alpha - j\beta = |F(j\omega)|\ e^{-j\phi}$$

where $$|F(j\omega)| = \sqrt{\alpha^2 + \beta^2} = M$$

and $$\phi = \tan^{-1}(\beta/\alpha).$$

Now we can write

$$y(t) = |F(j\omega)| \left[\frac{\exp(j(\omega t + \phi)) - \exp(-j(\omega t + \phi))}{2j}\right]$$

$$= |F(j\omega)| \sin(\omega t + \phi)$$

since $\sin\theta = (e^{j\theta} - e^{-j\theta})/2j.$

This proves the result stated above. It is now relatively straightforward to obtain the frequency response of a given transfer function. Graphically, this involves presentation of the relationship between the two dependent variables M and ϕ, and the independent variable ω. There are three possible ways of showing this relationship graphically: the Nyquist plot, the Bode plot and the Nichols plot.

The Nyquist plot is drawn on polar coordinates using the magnitude and phase of the response, with frequency as a parameter along the curve. The Bode plot is drawn as a pair of graphs on cartesian coordinates, with the phase and the logarithmic magnitude (dB) as the coordinates and the logarithmic frequency as the abscissa. Logarithmic magnitude (LM) is defined as

$$LM = 20\log_{10}|F(j\omega)| \text{ dB}.$$

The Nichols plot is drawn on cartesian coordinates, with the phase and the logarithmic magnitude as ordinate and abscissa respectively, and frequency as a parameter along the curve.

Some simple examples will be given to illustrate each of these plots.

EXAMPLE 5.14. Determine the frequency response of the first-order transfer function $3/(1+sT)$, and illustrate it using the Nyquist, Bode and Nichols plots.

Replacing s by $j\omega$ in the transfer function $F(s) = 3/(1 + sT)$ gives

$$F(j\omega) = 3/(1 + j\omega T)$$

Rationalising $F(j\omega)$:

$$\frac{3}{(1 + j\omega T)} \cdot \frac{(1 - j\omega T)}{(1 - j\omega T)} = \frac{3(1 - j\omega T)}{1 + \omega^2 T^2},$$

$$F(j\omega) = \frac{3}{1 + \omega^2 T^2} - \frac{j3\omega T}{1 + \omega^2 T^2}.$$

The magnitude M and phase ϕ are then given by:

$$M = |F(j\omega)| = \sqrt{(\text{Real part})^2 + (\text{Imag. part})^2}$$

$$= \left[\frac{9}{(1 + \omega^2 T^2)^2} + \frac{9\omega^2 T^2}{(1 + \omega^2 T^2)^2}\right]^{\frac{1}{2}}$$

$$= 3\left[\frac{1 + \omega^2T^2}{(1 + \omega^2T^2)^2}\right]^{\frac{1}{2}}$$

$$= 3/\sqrt{(1 + \omega^2T^2)}$$

and $$\phi = \underline{/F(j\omega)} = \text{arctan (Imag. part/Real part)}$$

$$= \text{arctan } (-\omega T).$$

Note that the gain factor of 3 appears in the expression for M but not in that for ϕ.

TABLE 5.2 Results for Example 5.14

ωT	$\log_{10} \omega T$	M	LM (dB)	ϕ (degrees)
0.001	-3.0	3.00	9.54	-0.057
0.01	-2.0	3.00	9.54	-0.573
0.02	-1.7	3.00	9.54	-1.15
0.05	-1.3	3.00	9.53	-2.86
0.10	-1.0	2.99	9.50	-5.71
0.20	-0.7	2.94	9.37	-11.3
0.50	-0.3	2.68	8.57	-26.6
1.0	0.0	2.12	6.53	-45.0
2.0	0.3	1.34	2.56	-63.4
5.0	0.7	0.588	-4.61	-78.7
10.0	1.0	0.299	-10.5	-84.3
20.0	1.3	0.150	-16.5	-87.1
50.0	1.7	0.060	-24.4	-88.9
100.0	2.0	0.030	-30.5	-89.4
1000.0	3.0	0.003	-50.5	-89.9

The logarithmic magnitude LM is given by

$$LM = 20 \log_{10} (3/\sqrt{(1 + \omega^2T^2)}) \text{ dB}$$

$$= 20 \log_{10}(3) - 10 \log(1 + \omega^2T^2)$$

since $$\log(A/B) = \log A - \log B$$

and $$\log \sqrt{A} = \frac{1}{2} \log A$$

Since the frequency ω always appears in the above expressions as a product with the time constant T, it is convenient to treat ωT as a single variable. It may be regarded as a normalised frequency. Table 5.2 shows the variation of M, LM and ϕ with ω. The three types of plot are shown in Fig. 5.33(a), (b) and (c). From

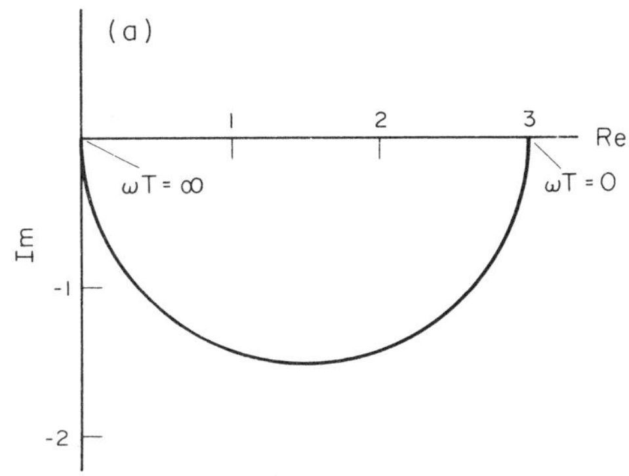

Fig. 5.33. (a) Nyquist plot for first-order transfer function, Example 5.14.

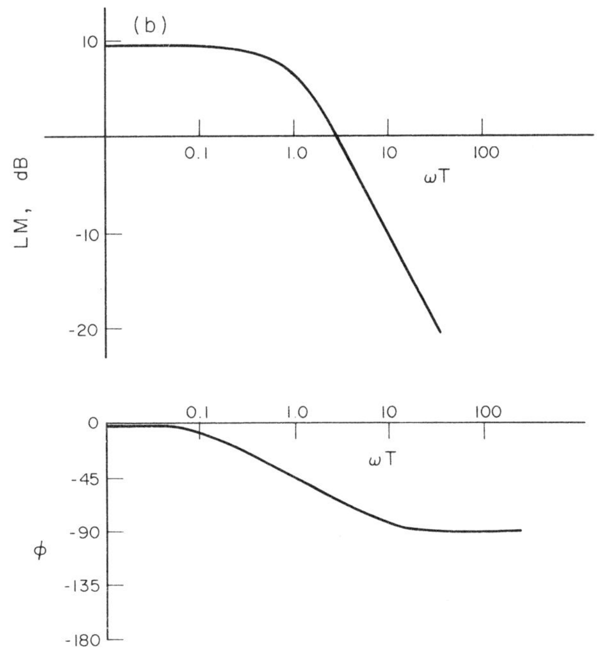

Fig. 5.33. (b) Bode plot for a first-order transfer function, Example 5.14.

these plots it can be seen that at low frequency the magnitude and phase are 3 and

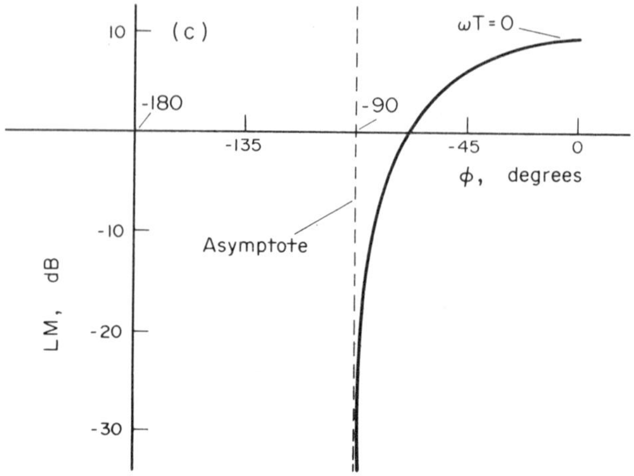

Fig. 5.33. (c) Nichols plot for a first-order transfer function, Example 5.14.

zero respectively. As frequency increases, the magnitude decreases continuously and the phase becomes more negative. At high frequency the magnitude approaches zero and the phase is asymptotic to -90°. Note that the magnitude is proportional to the gain constant, 3 in this case. Multiplying the gain constant by a factor A would multiply the scale of the Nyquist plot by A without changing the shape of the curve, and would add 20 log A decibels to the logarithmic magnitude in the Bode and Nichols plots, again without changing the shape of the curves.

EXAMPLE 5.15. Determine the frequency response of the second-order transfer function $\omega_n^2/(\omega_n^2 + 2\zeta\omega_n s + s^2)$, for $\zeta = 0.1$, 0.4, 0.6 and 0.9. Illustrate the response using Nyquist, Bode and Nichols plots.

For the transfer function

$$F(s) = \frac{\omega_n^2}{\omega_n^2 + 2\zeta\omega_n s + s^2},$$

replacing s by $j\omega$ gives

$$F(j\omega) = \frac{\omega_n^2}{\omega_n^2 + j2\zeta\omega_n\omega - \omega^2}$$

and simplifying by dividing numerator and denominator by ω_n^2 gives the dimensionless form:

$$F(j\omega) = \frac{1}{1 + j2\zeta r - r^2} = \frac{1}{(1 - r^2) + j2\zeta r}$$

where $r = \omega/\omega_n$ (dimensionless frequency).

Rationalising gives the form:

$$F(j\omega) = \frac{(1 - r^2)}{(1 - r^2)^2 + 4\zeta^2 r^2} - j\frac{2\zeta r}{(1 - r^2)^2 + 4\zeta^2 r^2}$$

and in magnitude and phase form:

$$|F(j\omega)| = 1/\sqrt{(1 - r^2)^2 + 4\zeta^2 r^2}$$

$$\phi = \arctan\,(-2\zeta r/(1 - r^2)).$$

The logarithmic magnitude is then:

$$\begin{aligned} LM &= 20\ \log|F(j\omega)| \\ &= 20\ \log(1/\sqrt{(1 - r^2)^2 + 4\zeta^2 r^2}\,) \\ &= -\ 10\ \log\,((1 - r^2)^2 + 4\zeta^2 r^2)\ \ \text{dB}. \end{aligned}$$

These responses are tabulated in Table 5.3, and shown in Figure 5.34 (a), (b) and (c). It can be seen that the behaviour at low frequency is similar to that of the first-order system in the previous example. As the frequency ratio "r" approaches unity the change in phase is much greater than previously, and the magnitude shows

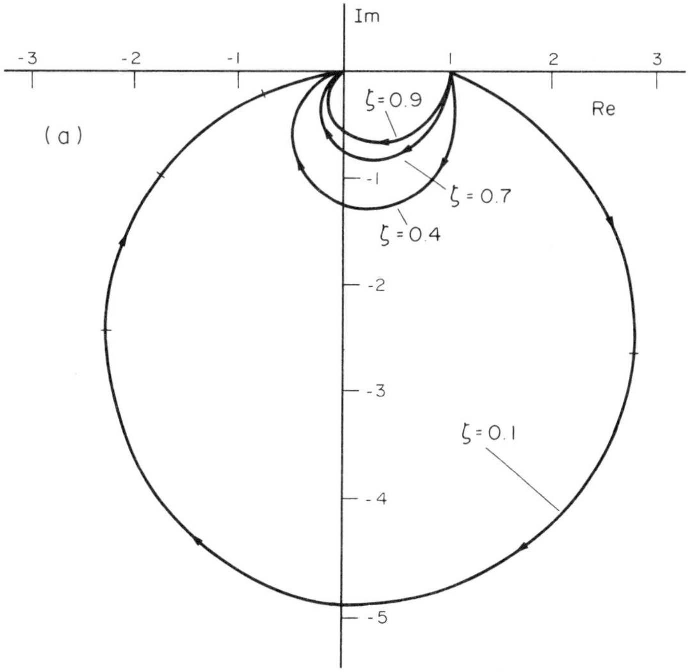

Fig. 5.34. (a) Nyquist plots for Example 5.15.

a resonance whose peak value depends on the value of the damping ratio ζ. The peak has a value of $(1/2\zeta)$ and occurs at a frequency $\omega_n\sqrt{1 - \zeta^2}$, which is close to the undamped natural frequency ω_n. At high frequency the phase approaches -180°, with

TABLE 5.3 Results for Example 5.15

Frequency ratio	Magnitude				Phase (degrees) (All values are negative)				Log magnitude (dB)			
	Damping ratio(ζ)				Damping ratio(ζ)				Damping ratio(ζ)			
	0.1	0.4	0.7	0.9	0.1	0.4	0.7	0.9	0.1	0.4	0.7	0.9
0.01	1.00	1.00	1.00	1.00	0.11	0.45	0.80	1.00	0.0	0.0	0.0	-0.0
0.10	1.01	1.00	1.00	0.99	1.16	4.6	8.1	10.3	0.09	0.06	0.0	-0.1
0.20	1.04	1.03	1.00	0.98	2.38	9.5	16.0	21.0	0.35	0.24	0.0	-0.2
0.50	1.32	1.18	0.98	0.85	7.60	28.0	43.0	50.0	2.42	1.40	-0.2	-1.4
0.75	2.16	1.35	0.88	0.70	18.8	54.0	67.0	72.0	6.70	2.60	-1.1	-3.0
1.0	5.00	1.25	0.71	0.56	90.0	90.0	90.0	90.0	14.0	1.90	-2.9	-5.1
1.5	0.78	1.58	0.41	0.34	166	136	121	115	-2.2	-4.8	-7.8	-9.5
2.0	0.33	0.29	0.24	0.21	172	152	137	130	-9.6	-10.6	-12.3	-13.4
3.0	0.12	0.12	0.11	0.10	176	163	152	146	-18	-18	-19	-20
5.0	0.04	0.04	0.04	0.04	178	171	164	159	-28	-28	-28	-28
10.0	0.01	0.01	0.01	0.01	179	175	172	170	-40	-40	-40	-40
100	← 1×10^{-4} →				180	180	179	179	-80	-80	-80	-80

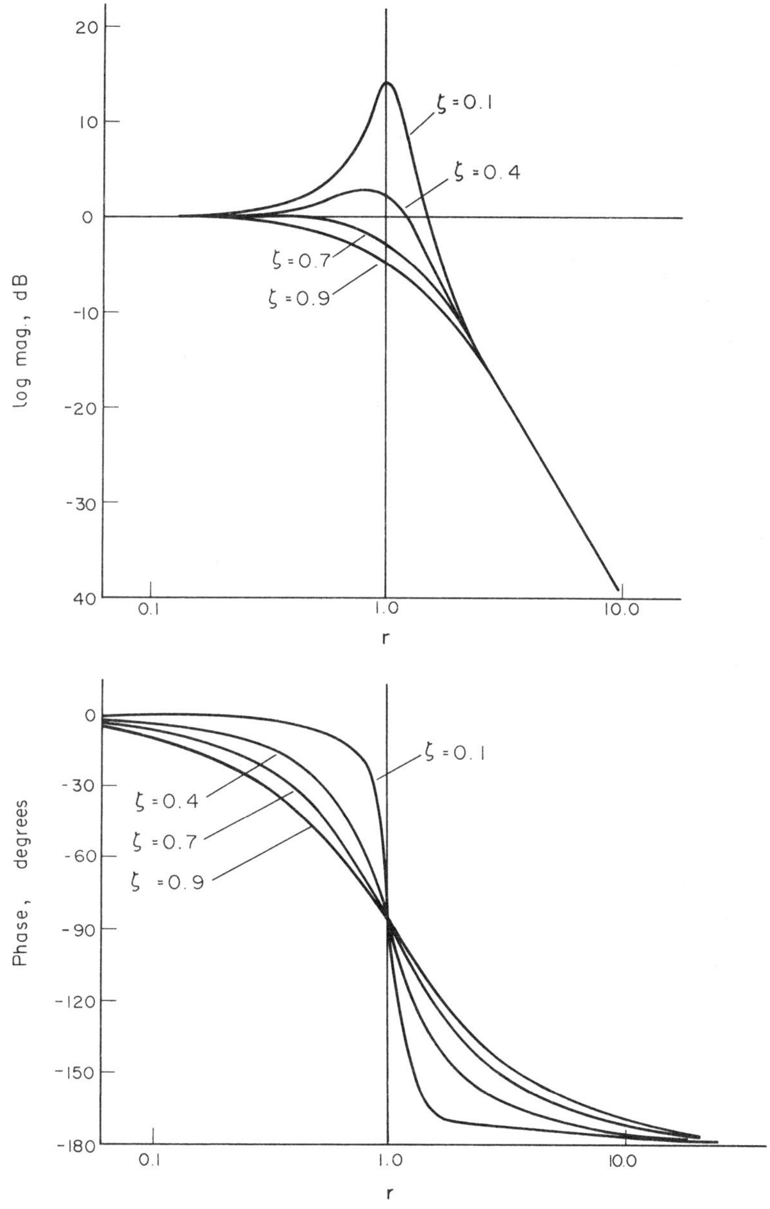

Fig. 5.34. (b) Bode plots for Example 5.15.

magnitude decreasing at -40 dB per decade instead of -20 dB/decade as in the first-order case.

Rules for frequency response plots. By examination of the transfer function and application of simple rules it is possible to construct the general form of the frequency response, particularly for polar plots.

Considering the general linear transfer function

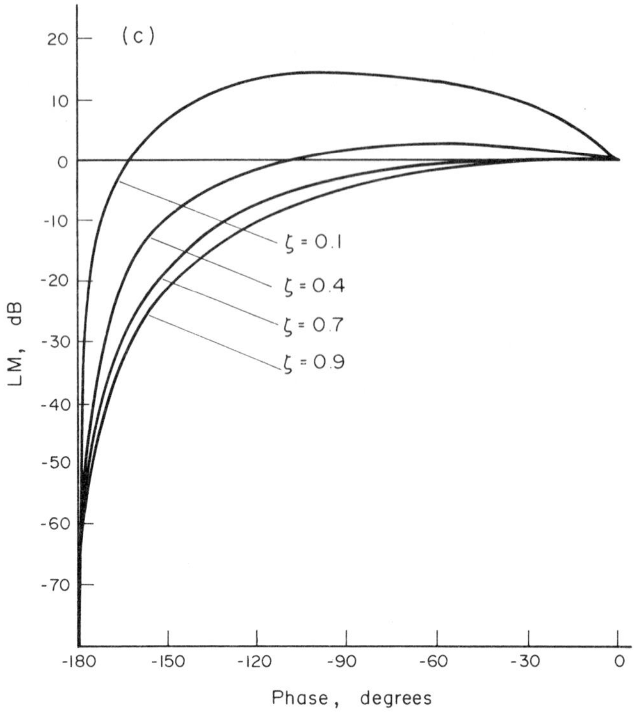

Fig. 5.34. (c) Nichols plot for Example 5.15.

$$F(s) = \frac{K(s - a_1)(s - a_2)(s - a_3) \ldots (s - a_m)}{s^r(s - b_1)(s - b_2)(s - b_3) \ldots (s - b_n)} .$$

The rank is $r + n - m$.

The type is r.

The order is $r + n$.

We have rules as follows, which can generally be deduced from the initial value and final-value theorems.

1. Magnitude at low frequency

   For a type 0 system $M = K$ at low frequency. For a type 1 (or higher) system, $M = \infty$ at low frequency.

2. Magnitude at high frequency

   For a transfer function representing any real physical system, the rank will always be greater than zero, i.e. $(r + n) > m$, hence at high frequency the magnitude will always approach zero.

3. Phase at low frequency

$$\phi_{\omega \to 0} = (-90^o) \times (\text{function type}).$$

4. Phase at high frequency

$$\phi_{\omega \to 0} = (-90^o) \times (\text{function rank}).$$

The frequency response of functions containing pure delay terms such as exp(-Ts) can be obtained in a straightforward manner, because the function exp(jωT) can be written in explicit, closed form instead of using the series approach given in Section 5.9.3.

Since
$$\exp(-j\omega t) = \cos \omega T - j \sin \omega T$$

and
$$|\exp(-j\omega T)| = \sqrt{\cos^2 \omega T + \sin^2 \omega T} = 1,$$

$$\underline{/\exp(-j\omega T)} = \arctan(-\sin \omega T/\cos \omega T)$$
$$= \arctan(-\tan \omega T)$$
$$= -\omega T$$

This is a term whose magnitude is always unity but whose contribution to the phase increases without limit, as the frequency increases.

EXAMPLE 5.16. Repeat the problem of Example 5.14 for a transfer function

$$F(s) = 3\cdot\exp(-s)/(1 + s).$$

Hence
$$F(j\omega) = 3 \exp(-j\omega)/(1 + j\omega).$$

The expressions for magnitude and log magnitude will be the same as in Example 5.14, but the phase will be more negative by ω radians or 57.3ω degrees.

$$M = 3/\sqrt{1 + \omega^2},$$

$$LM = 20 \log M$$
$$= 20 \log 3 - 10 \log (1 + \omega^2)$$
$$= 9.54 - 10 \log (1 + \omega^2) \quad \text{dB}$$

$$\phi = \arctan(-\omega) - 57.3\omega \text{ degrees.}$$

This is tabulated in Table 5.4 and shown in Fig. 5.35 (a), (b) and (c). It is apparent that the presence of the exponential term has radically changed the shape of the plots.

5.10.1 Asymptotic Approximations in Bode Plots

In Examples 5.14 and 5.15, the behaviour of both the log-magnitude and the phase plots of the Bode diagrams was asymptotic to straight lines. This fact leads to a simple method of constructing approximate Bode plots without specific calculations.

Consider the general linear transfer function F(s) in the following form:

$$F(s) = \frac{K(1 + sT_1)(1 + sT_2)(1 + sT_3) \ldots (1 + sT_m)}{s^r(1 + sT_1')(1 + sT_2') \ldots (1 + sT_n')(1 + (2\zeta/\omega_n)s + (1/\omega_n)^2 s^2)} \quad (5.43)$$

where the T_1 to T_m and the T_1' to T_n' are real positive time constants, and there is a quadratic term in the denominator.

TABLE 5.4 Results for Example 5.16

ω (radian/sec)	$\log_{10}\omega$	M	LM (dB)	ϕ (degrees)
0.001	-3.0	3.00	9.54	-0.114
0.01	-2.0	3.00	9.54	-1.15
0.02	-1.7	3.00	9.54	-2.30
0.05	-1.3	3.00	9.53	-5.73
0.1	-1.0	2.99	9.50	-11.4
0.2	-0.7	2.94	9.37	-22.8
0.5	-0.3	2.68	8.57	-55.2
1.0	0.0	2.12	6.53	-102
2.0	0.3	1.34	2.56	-108
5.0	0.7	0.588	-4.6	-365
10.0	1.0	0.299	-10.5	-657
20.0	1.3	0.150	-16.5	-1,230
50.0	1.7	0.060	-24.4	-2,950
100.0	2.0	0.030	-30.5	-5,820
1000.0	3.0	0.003	-50.5	-57,400

We obtain the expressions for the Bode plot as follows:

$$\text{Logarithmic Magnitude, LM} = 20 \log |F(j\omega)|$$

$$= 20 \log|K| + 20 \log |1 + j\omega T_1| + 20 \log |1 + j\omega T_2|$$

$$+ 20 \log |1 + j\omega T_3| + \ldots + 20 \log |1 + j\omega T_m|$$

$$- 20 \log |(j\omega)^r| - 20 \log |1 + j\omega T_1'| - 20 \log |1 + j\omega T_2'| \ldots$$

$$- 20 \log |1 + (2\zeta/\omega_n)j\omega + (1/\omega_n)^2(j\omega)^2|$$

Considering each term separately:

1. $20 \log|K| = 20 \log (K)$ since K is a real positive constant. This gives a constant contribution to the LM, independent of frequency. Changes in K move the whole of the LM plot up or down, parallel to the LM axis.

2. For a term of the form $20 \log |1 + j\omega T|$, we have

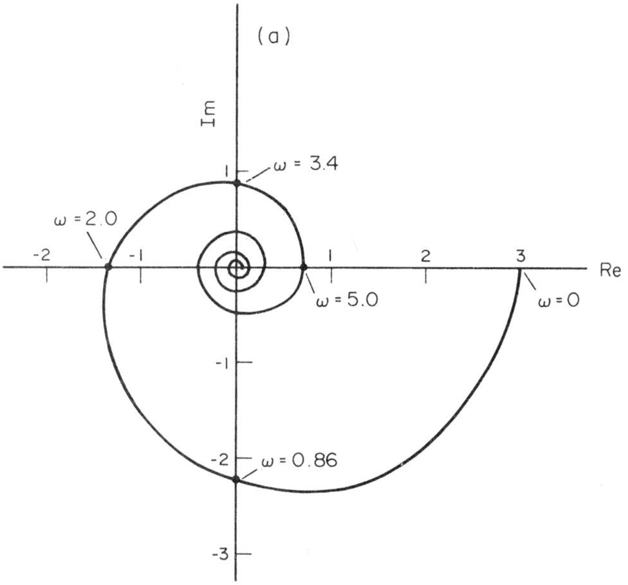

Fig. 5.35. (a) Nyquist plot for Example 5.16.

$$|1 + j\omega T| = \sqrt{1 + \omega^2 T^2} \ ,$$

$$20 \log |1 + j\omega T| = 20 \log \sqrt{1 + \omega^2 T^2}$$

$$= 10 \log (1 + \omega^2 T^2) \quad \text{dB}.$$

For low frequencies ($\omega \ll 1/T$) the above expression approximates to

$$10 \log(1) = 0 \text{ dB}.$$

For high frequencies ($\omega \gg 1/T$) the same expression will approximate to

$$10 \log (\omega^2 T^2) = 20 \log (\omega T) \quad \text{dB},$$

which undergoes a change of ± 20 dB for each decade (factor of 10) change in frequency. This gives a straight line with a slope of +20 dB/decade for terms in the numerator and -20 dB/decade for terms in the denominator. This is shown in Fig. 5.36(a).

3. For the s^r term we have

$$|(j\omega)^r| = \omega^r$$

and
$$-20 \log (\omega^r) = -20r \cdot \log(\omega) \quad \text{dB}.$$

Hence, for an s^r term in the denominator, the LM plot is a straight line of slope -20·r dB/decade, passing through 0 dB at a frequency $\omega = 1$ rad/sec. It is unusual for r to exceed 3. This is illustrated in Fig. 5.36(b).

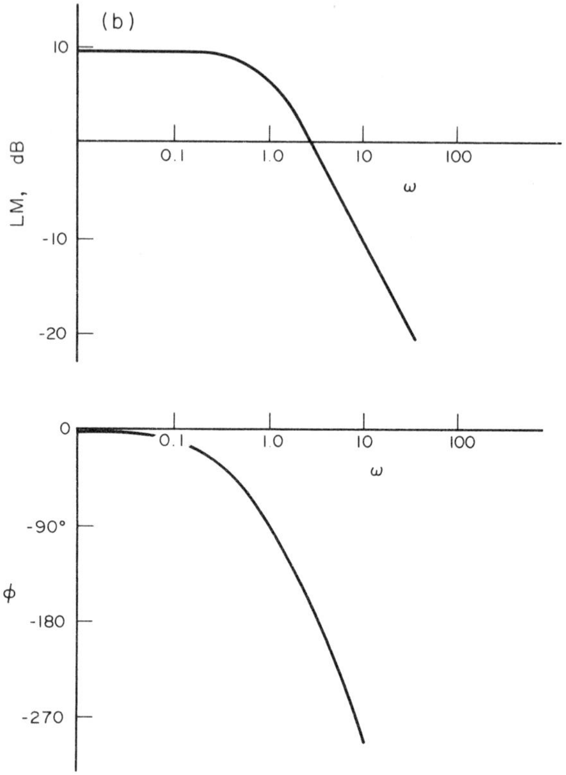

Fig. 5.35. (b) Bode plot for Example 5.16.

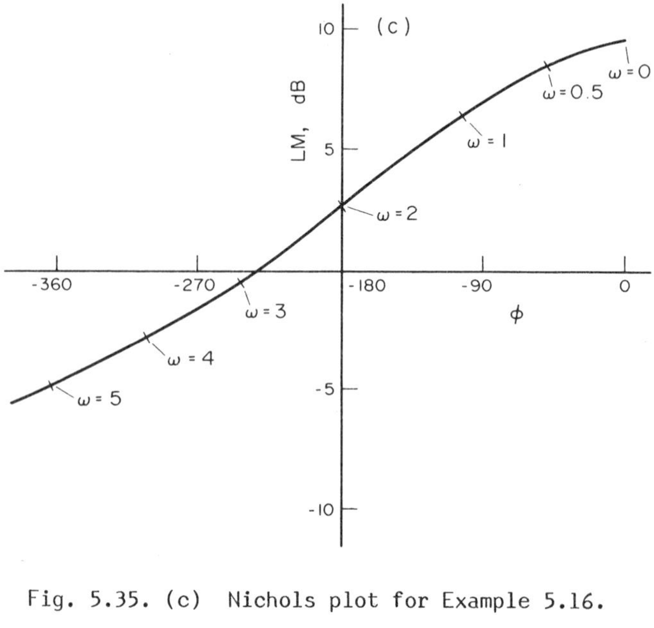

Fig. 5.35. (c) Nichols plot for Example 5.16.

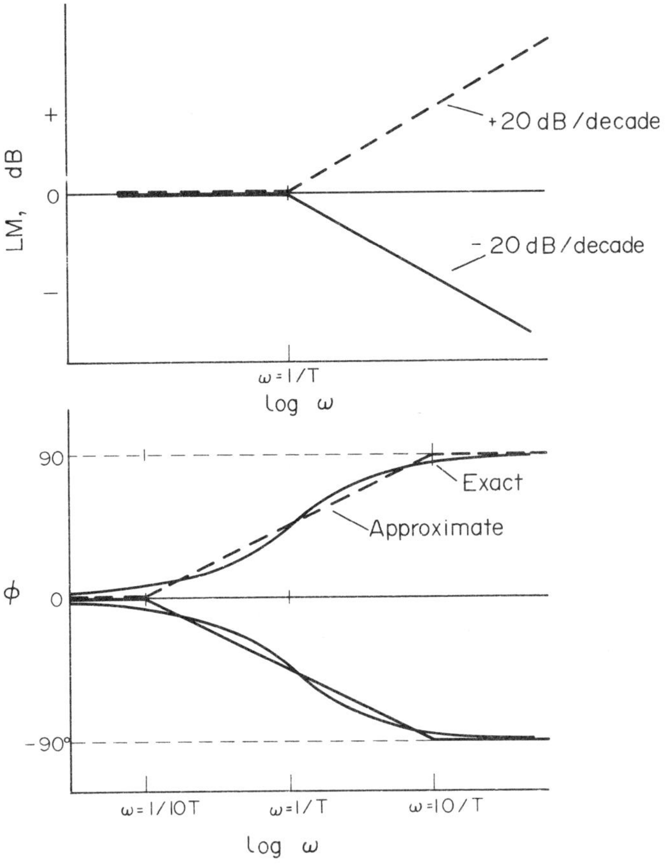

Fig. 5.36. (a) Asymptotic approximations for $(1+Ts)$ term.
----- Numerator
—— Denominator

4. For the quadratic term in the denominator, the behaviour of the LM in the region $\omega \simeq \omega_n$ depends strongly on the value of the damping ratio ζ. For $\zeta \geqslant 1$ we can factorise into two first-order terms with real roots, which can be treated as outlined above. For $\zeta \geqslant 1$ the entire quadratic can be plotted.

$$LM = -20 \log \left|1 + (2\zeta/\omega_n)j\omega + (1/\omega_n)^2(j\omega)^2\right|$$

$$= -20 \log \sqrt{(1 - \omega^2/\omega_n^2)^2 + (2\omega/\omega_n)^2}.$$

Now considering the three frequency ranges of interest:

$$\omega \ll \omega_n: \; LM \simeq -20 \log(1) = 0 \text{ dB},$$

$$\omega \simeq \omega_n: \; LM \simeq -20 \log(2\zeta) \quad \text{dB},$$

$$\omega \gg \omega_n: \; LM \simeq -20 \log(\omega^2/\omega_n^2)$$

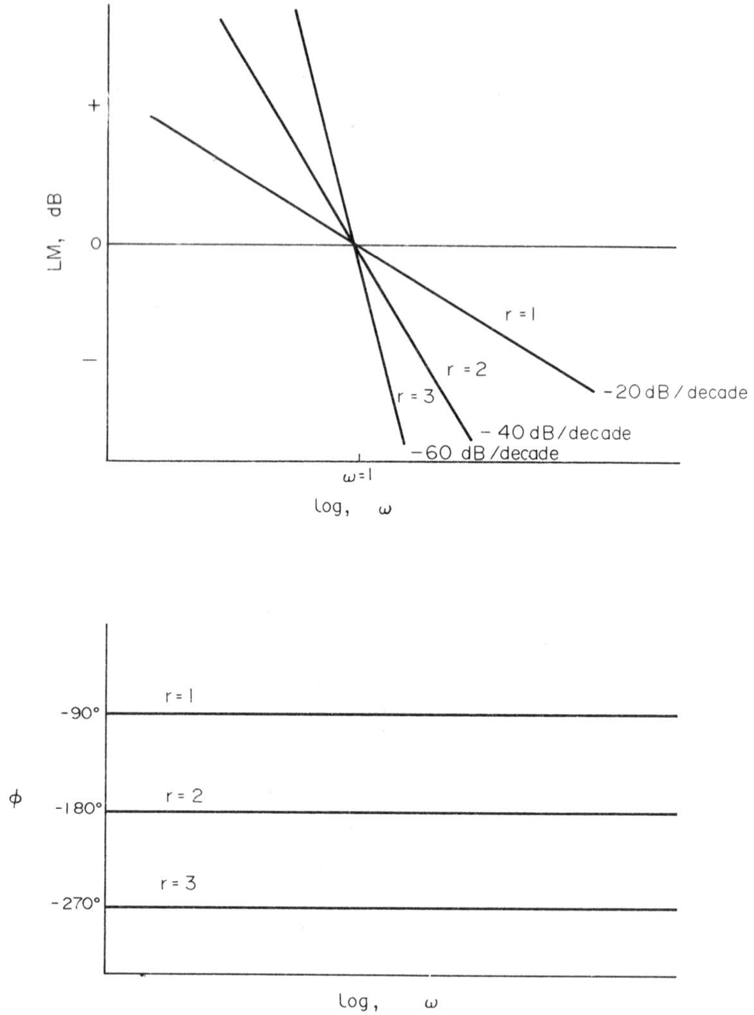

Fig. 5.36. (b) Asymptotic approximations for s^r term.

$$= -40 \log(\omega/\omega_n) \quad dB$$

since the second term in brackets above will become negligible and $(1 - \omega^2/\omega_n^2)^2 \approx \omega^4/\omega_n^4$.

From this we have a constant horizontal straight line at 0 dB for low frequencies, a resonant peak at $\omega = \omega_n \sqrt{1 - 2\zeta^2}$ if $\zeta < 0.5$, and a straight line asymptote with a slope of -40 dB/decade for $\omega >> \omega_n$, high frequencies, as illustrated in Fig. 5.36(c).

Now returning to equation (5.43) to consider the phase response, we have

$$\phi = \underline{/F(j\omega)}$$

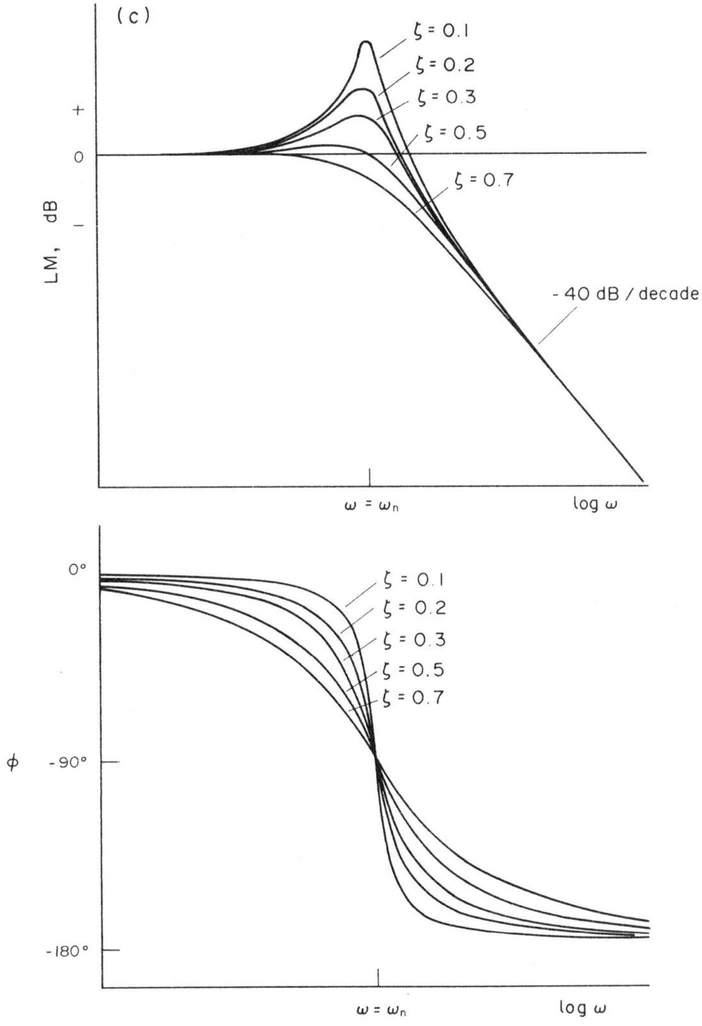

Fig. 5.36. (c) Bode plot for quadratic term in denominator.

$$= \underline{/K} + \underline{/1 + j\omega T_1} + \underline{/1 + j\omega T_2} + \ldots + \underline{/1 + j\omega T_m}$$

$$- \underline{/(j\omega)^r} - \underline{/1 + j\omega T_1'} - \underline{/1 + j\omega T_2'} - \ldots - \underline{/1 + j\omega T_n'}$$

$$- \underline{/1 + (2\zeta/\omega_n)j\omega + (1/\omega_n)^2(j\omega)^2}$$

Again taking each term separately:

1. $\underline{/K}$ is zero since K is a real positive constant, making no contribution to the phase response. This is important in that the gain constant determines the size or scale of the response plot, but not its shape.

2. The terms of form $(1 + j\omega T)$ have a phase contribution arctan (ωT) for numerator terms or arctan $(-\omega T)$ for denominator terms. The phase thus runs from zero at low frequency $(\omega << 1/T)$ to $\pm 45^o$ when $\omega = 1/T$ and to $\pm 90^o$ at high frequency $(\omega >> 1/T)$. This is shown in Fig. 5.36(a), which also illustrates a straight-line approximation for the phase plot. The maximum error using this approximation is less than 6°.

3. For the s^r term in the denominator we have

$$- \underline{/(j\omega)^r} = - \underline{/0 + j\omega} = -90^o \text{ for } r = 1$$

$$\text{or} \quad = - \underline{/-\omega^2 + 0} = -180^o \text{ for } r = 2$$

$$\text{or} \quad = - \underline{/0 - j\omega^3} = -270^o \text{ for } r = 3$$

In this case the phase plot coincides with the asymptote, as shown in Fig. 5.36(b).

4. The phase of the quadratic term in the denominator is given by

$$\phi = - \arctan\left(\frac{2\zeta\omega/\omega_n}{1-\omega^2/\omega_n^{\ 2}}\right).$$

For the three frequency ranges of interest, we have:

$$\omega << \omega_n: \ \phi \approx - \arctan (0) \qquad = \quad 0^o,$$

$$\omega \approx \omega_n: \ \phi \approx - \arctan (\infty) \qquad = \ -90^o,$$

$$\omega >> \omega_n: \ \phi \approx - \arctan (-2\zeta\omega_n/\omega) \approx -180^o.$$

The behaviour at very low and very high frequencies is straightforward: unfortunately there is no simple straight-line approximate construction for the behaviour at mid-frequencies, and it becomes necessary to plot the curve by calculating and plotting points explicitly.

In order to illustrate these approximate plots, an example will be shown of their use.

<u>EXAMPLE 5.17</u>. Use asymptotic approximations to draw the Bode plot for the transfer function

$$F(s) = 10\ (s + 3)/s(s + 1)(2 + s + s^2).$$

Rearranging the function to obtain the form of equation (5.43) gives

$$F(s) = \frac{15\ (1 + 0.33s)}{s(1 + s)(1 + 0.5s + 0.5s^2)}$$

The gain factor 15 gives a LM contribution of 20 log(15) = + 23.5 dB.

The numerator term $(1 + 0.33s)$ has a "corner" frequency $\omega = 1/0.33 = 3$ rad/sec.

The denominator term $(1 + s)$ has a "corner" frequency $\omega = 1$ rad/sec.

The denominator term "s" is as in Fig. 5.36(b) for $n = 1$.

For the quadratic term in the denominator we have:

$$2\zeta/\omega_n = 0.5 \text{ and } 1/\omega_n^2 = 0.5,$$

hence $$\omega_n = \sqrt{2} \text{ and } \zeta = 0.354.$$

The response of these individual terms and their sum are shown in Fig. 5.37 (a)

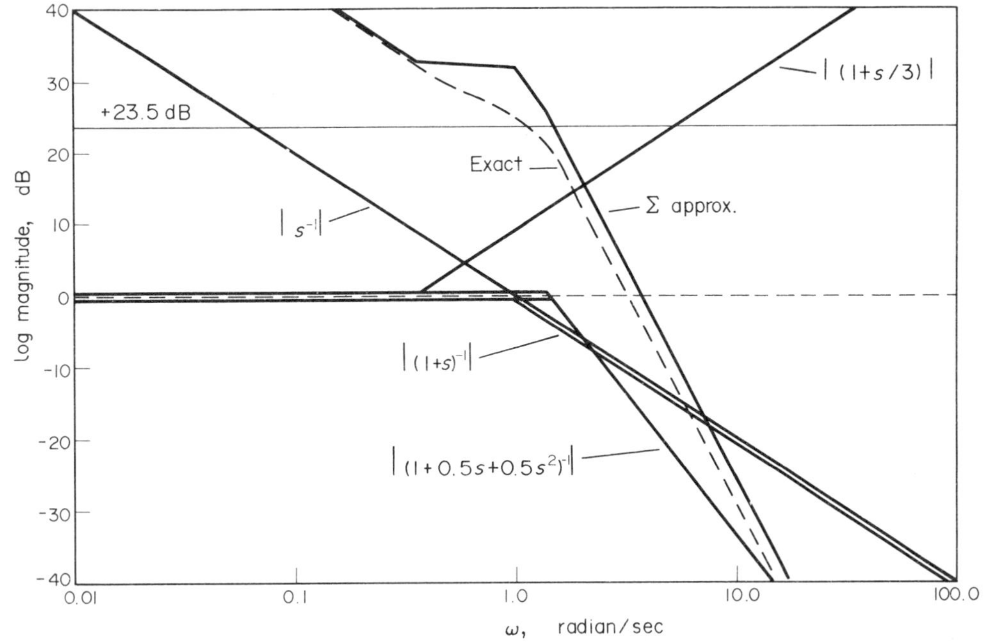

Fig. 5.37. (a) Log magnitude plots for Example 5.17.

and (b) The exact response is also shown and it can be seen that the errors involved in the asymptotic plot are quite small. This is a quick and reliable method of finding frequency responses, and involves very little calculation.

5.11 THE NYQUIST CRITERION

So far we have considered the frequency response of transfer functions for its own sake: there has been no conclusion to draw about the significance of a given type of frequency response. In order to give some meaning to the work of the previous sections, we now deal with the stability criterion which is applicable to frequency response data: the Nyquist criterion. As a preliminary it is necessary to deal with some relevant concepts.

1. The complex function as a mapping rule

A function $f(s)$ of the complex variable $s(\sigma + j\omega)$ can be considered as a rule for generating a new complex number $f(s)$ from the complex number s. Figure 5.38 illustrates this, and shows how a contour AB in the s-plane maps into a corresponding contour A'B' in the $f(s)$-plane. In particular, for functions of the type with which we are concerned, the mapping is *conformal*, that is, a

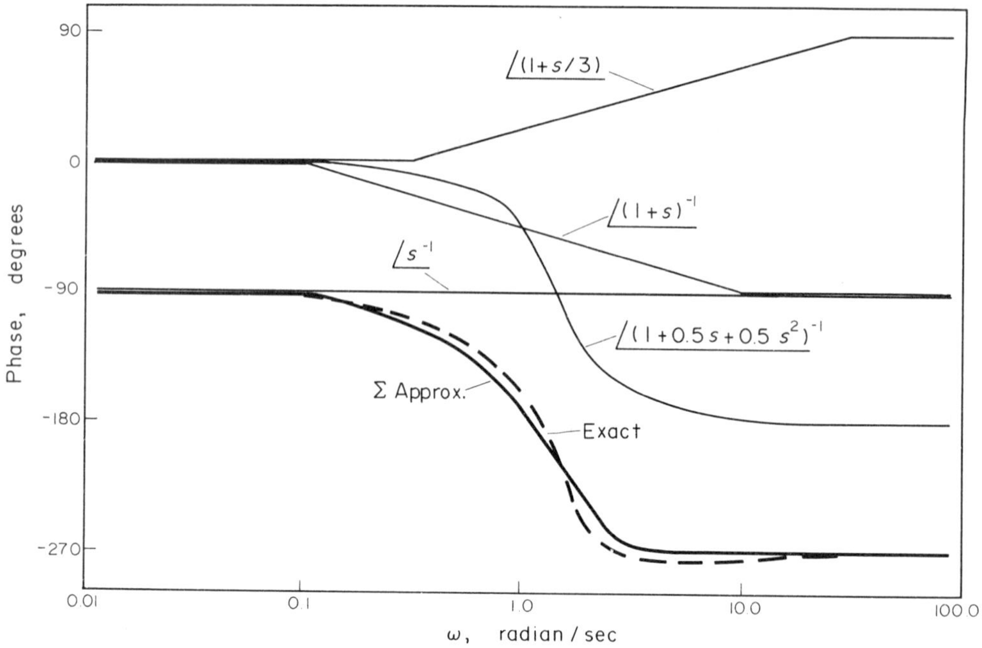

Fig. 5.37. (b) Phase plots for Example 5.17.

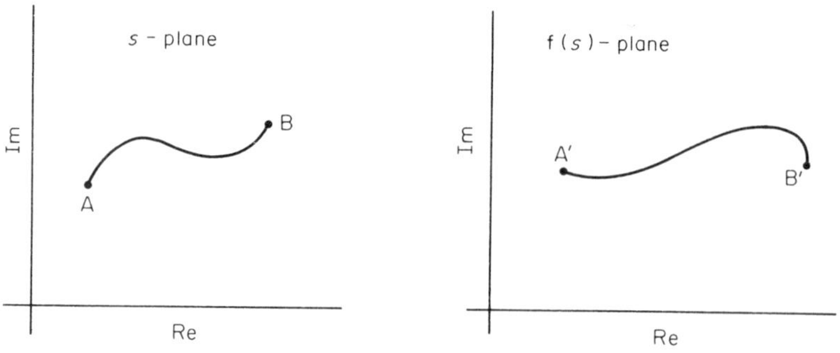

Fig. 5.38. Conformal mapping from the s-plane to the $f(s)$-plane.

closed contour in the s-plane will map into a closed contour (in general of different shape) in the $f(s)$-plane.

2. The frequency response of the characteristic function considered as a mapping

Applying the above concept to the characteristic function $1 + G(s)H(s)$, we can choose a contour in the s-plane which coincides with the positive imaginary axis, i.e. $s = +j\omega$. Allowing ω to vary from zero to positive infinity will give the mapping of the imaginary axis through the function $1 + G(j\omega)H(j\omega)$ into the $1 + G(j\omega)H(j\omega)$ plane, which corresponds to the frequency response of the characteristic function for positive values of frequency ω. For the negative half of the imaginary axis in the s-plane, we obtain the function $1 + G(-j\omega)H(-j\omega)$ which is the complex conjugate of the previous function, giving the mirror image of the previous contour, as shown in Fig. 5.39.

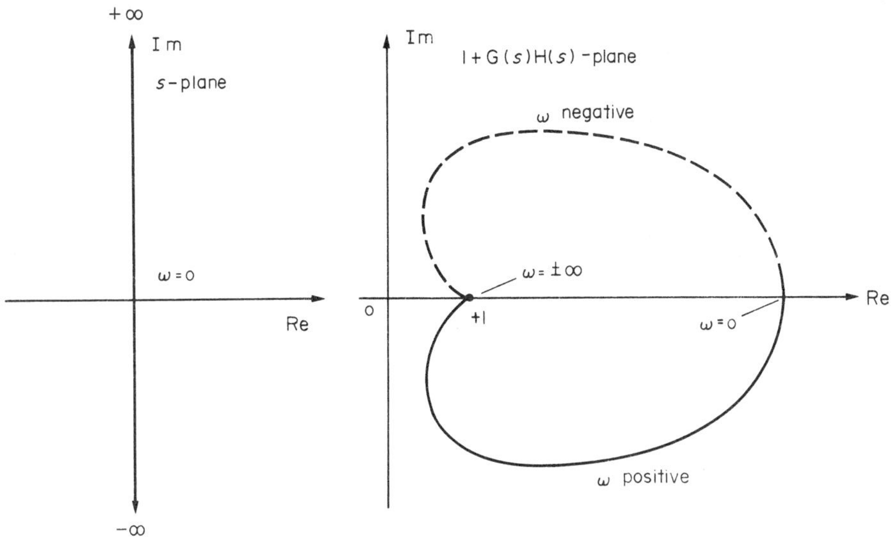

Fig. 5.39. Frequency response of the characteristic function.

3. The encirclement theorem

> This states that the net change in angle produced by traversing a closed contour is equal to $2\pi(P - Z)$ radians, where P and Z are the numbers of poles and zeros respectively enclosed within the contour. For a real physical system $P > Z$ always.

This is illustrated in Fig. 5.40, which shows two closed contours C_1 and C_2, one of which does enclose a pole and the other does not. Starting at the point A on C_1, the vector P_1A will undergo one complete rotation, and the vector P_2A will undergo zero net rotation as the contour is traversed completely. There will thus be a net change in angle of 2π radians for C_1, corresponding to one pole being enclosed. Similarly, for the contour C_2, the vectors P_1B and P_2B will both experience zero net rotation as B traverses the contour completely. This corresponds to no poles being enclosed by the contour C_2.

The Nyquist criterion involves a combination of the ideas outlined above, together with the result shown in Section 5.7, where it was shown that the stability of the closed-loop control system is determined by the location of the poles of the closed-loop transfer function, which are the roots of the characteristic equation $1 + G(s)H(s) = 0$. If any root lies in the right-hand half-plane then the system will be unstable. If any root lies on the imaginary axis (the borderline between RH and LH half-planes) then the system will exhibit continuous oscillations. This is on the boundary between stable and unstable response, although technically it would be termed unstable.

If we take as our contour in the s-plane, the path which includes the positive and negative halves of the imaginary axis, with the ends joined by a semicircle of infinite radius, the contour will be as shown in Fig. 5.41(a). This is often known as the D-contour. This will map into the frequency response of the characteristic function, shown in the $1 + G(s)H(s)$ plane in Fig. 5.41(b). We now

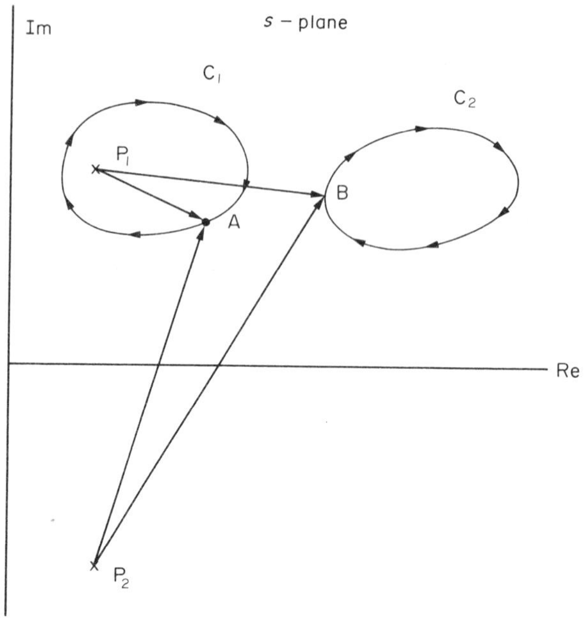

Fig. 5.40. The encirclement theorem.

employ the encirclement theorem to test whether any roots of $1 + G(s)H(s)$ are enclosed by the D-contour, by observing the net change in angle experienced by the characteristic function $1 + G(j\omega)H(j\omega)$.

Consider a vector from the origin in Fig. 5.41(b) to the point A. As the frequency increases positively the vector will move round clockwise passing through the point B until at very high frequency it reaches the point C, which is the point +1 on the real axis. The whole of the semicircular part of the D-contour maps into this point. As the frequency reduces from negative infinity, the vector will traverse the mirror-image, through the point D and back to the starting-point A as ω approaches zero. The net change in angle of this vector as its end traverses ABCD is zero, indicating that that particular characteristic function has *no* roots in the RH half of the s-plane and is therefore stable.

If we apply the same argument to a vector with one end at the origin and the other end traversing A'B'CD', in that case the net change in angle is 4π radians indicating that that function in unstable. The difference between the two contours is that A'B'CD' includes the origin of $1 + G(s)H(s)$ within it whereas ABCD does not enclose the origin.

In the application of the criterion there are two small modifications made to the procedure outlined above. Firstly, it is unnecessary to draw both the upper and the lower halves of the curves shown in Fig. 5.41(b), since they are mirror images of each other. Secondly, instead of plotting the frequency response of the characteristic function $1 + G(s)H(s)$, the open-loop transfer function $G(s)H(s)$ only is used, so that the whole curve is shifted to the left by unity, and the origin is replaced by the point -1, which is called the critical point. When these modifications are made to the curves of Fig. 5.41(b) we obtain the curves shown in Fig. 5.41(c).

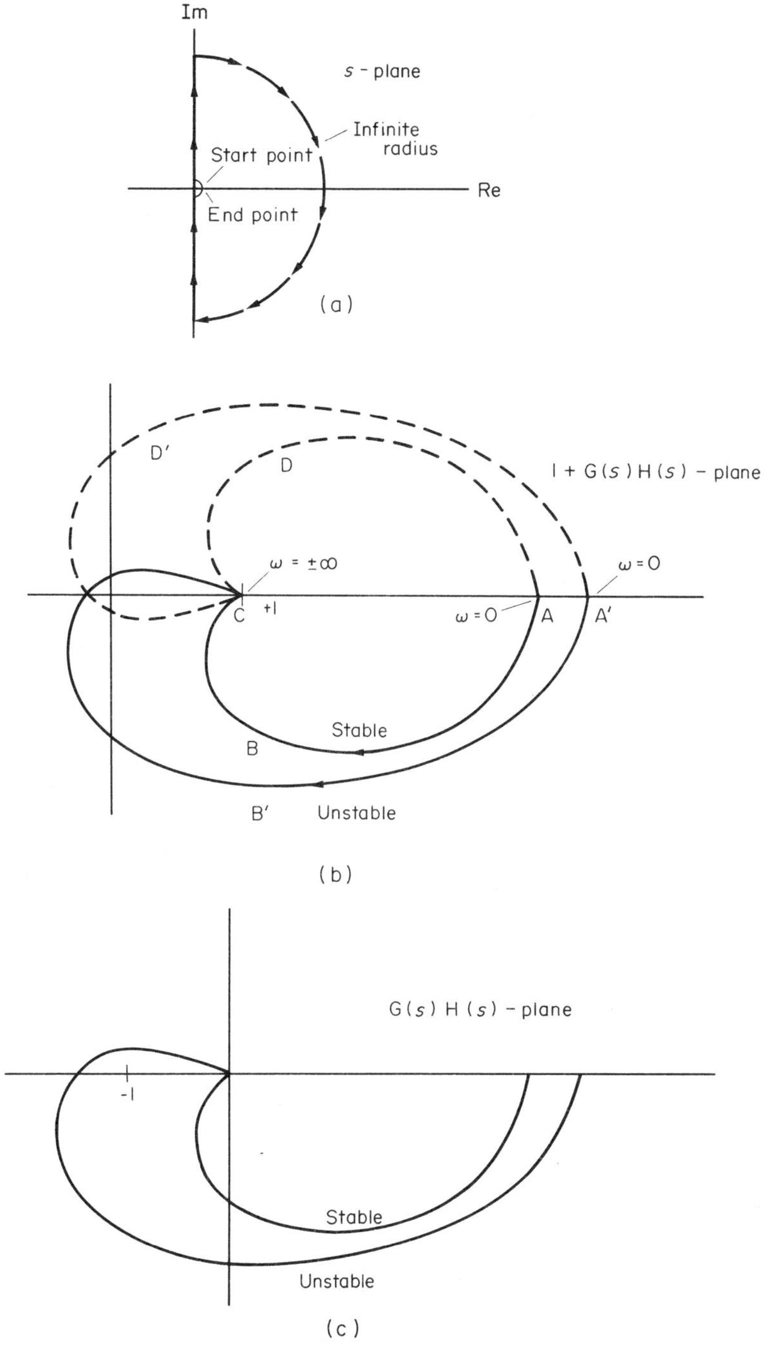

Fig. 5.41. Stable and unstable system plots in the Nyquist criterion.

5.11.1 The Nyquist Criterion using the Bode and Nichols Plots

In the previous section it was shown that the decision about the stability of the system can be expressed as a decision about its open-loop transfer function. If the magnitude of the frequency response of the open-loop transfer function is greater than unity when the phase lag is 180° then the system is unstable. This criterion can be illustrated in the Bode and Nichols plots as well as in the polar plot. Figure 5.42 shows how an increase in the gain constant moves the log-

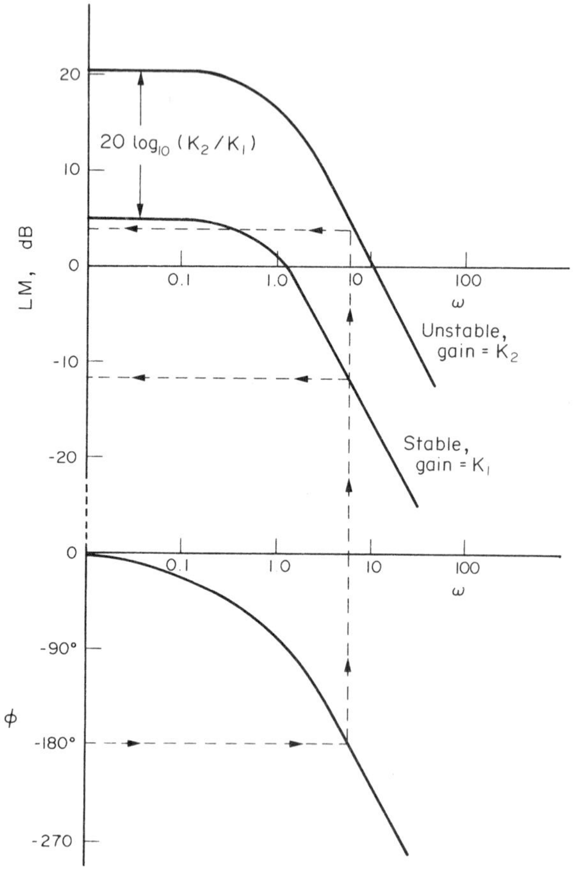

Fig. 5.42. Nyquist criterion in Bode plots - effects of gain change.

magnitude plot vertically and alters a stable system to an unstable one. There is no single critical point here; it is necessary to identify the frequency at which the phase plot reaches -180° and then determine whether the log-magnitude plot is positive or negative at this frequency. To facilitate this decision, the two curves of the Bode diagram are sometimes drawn superimposed, with the scales chosen so that the 0 dB line on the log-magnitude scale coincides with the -180° line on the phase scale. If the two curves cross above this line then instability is indicated: a crossing below the line indicates a stable system. This is shown in Fig. 5.43.

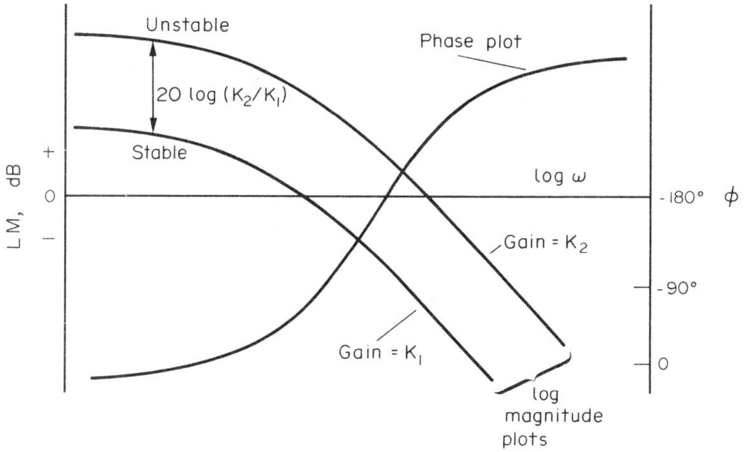

Fig. 5.43. Alternative form of Bode plot.

The Nichols plot can be used in an exactly similar way to determine stability. The origin of axes is chosen to be the combination of 0 dB, -180° so that if the curve crosses the log-magnitude axis above the origin an unstable system is indicated, and vice versa. This is illustrated in Fig. 5.44.

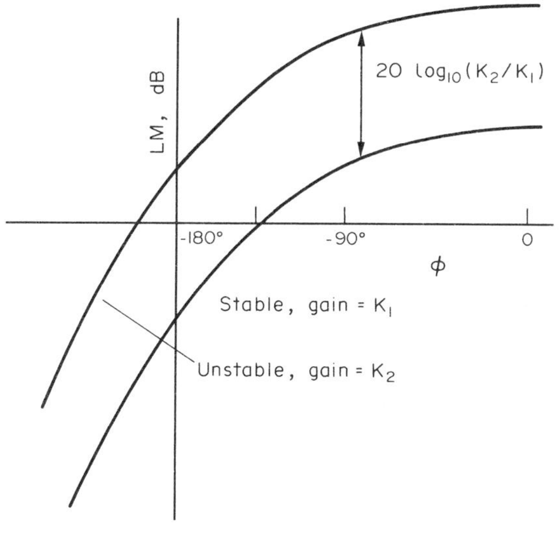

Fig. 5.44. Nyquist criterion using Nichols plot.

5.11.2 Phase and Gain Margins

The Nyquist criterion shares with the Root Locus method the ability to indicate how near a system is to being unstable. This is normally apparent from inspection of the curves, but it can be quantified by the use of the concepts of Phase Margin (PM) and Gain Margin (GM). They are defined as follows for a stable system:

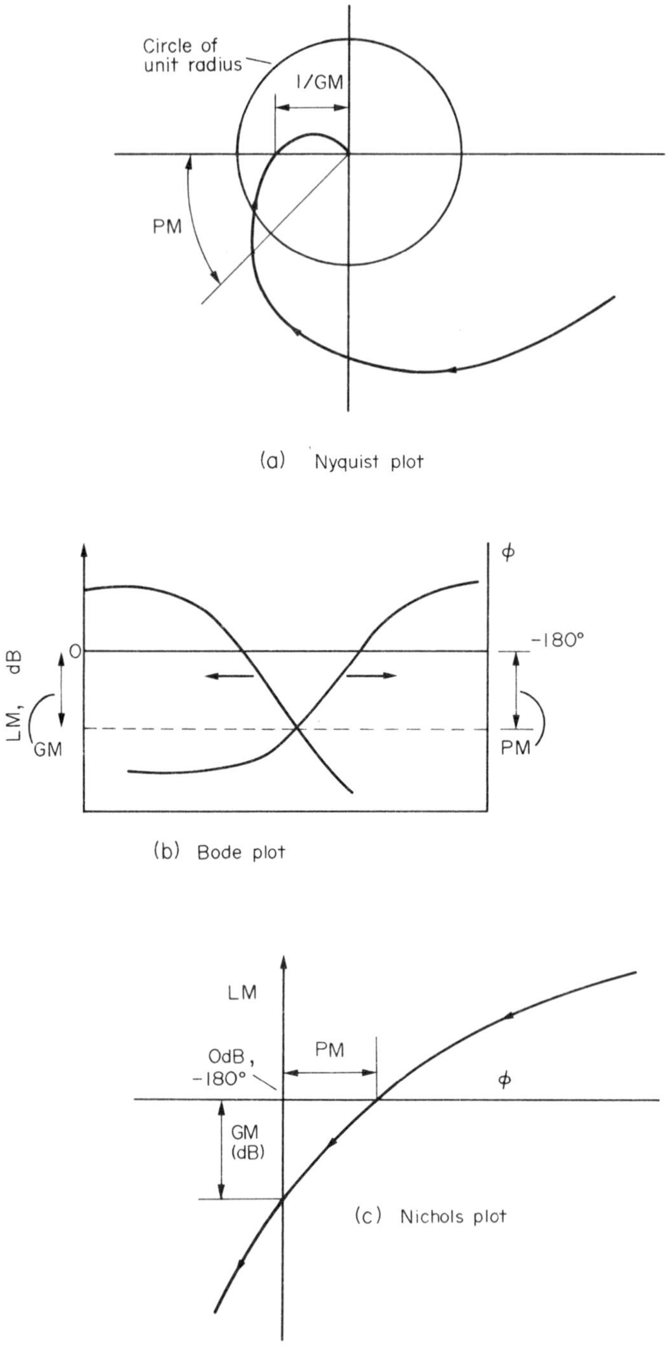

Fig. 5.45. Phase and gain margins in the three types of plots.

Phase Margin is the difference between -180° and the OLTF phase when the OLTF magnitude is unity.

Gain Margin is the number of decibels to be added to the log-magnitude when the phase is -180° to make the log-magnitude equal to zero.

These definitions are illustrated in Fig. 5.45 (a), (b) and (c).

Although it is obvious that the larger the values of PM and GM the further the system will be from instability, it is not possible to lay down universal rules for acceptable values of PM and GM. As a rough guide, a system with the combination of 5 dB GM and 40° PM will generally be acceptable. It is important, however, to note that *both* requirements must be satisfied: it is quite possible to have systems which are close to instability but which have acceptable values of either PM or GM. Figure 5.46 shows the polar plots of systems which exhibit in one case a large value of GM and in the other case a large value of PM, but which are both rather close to instability.

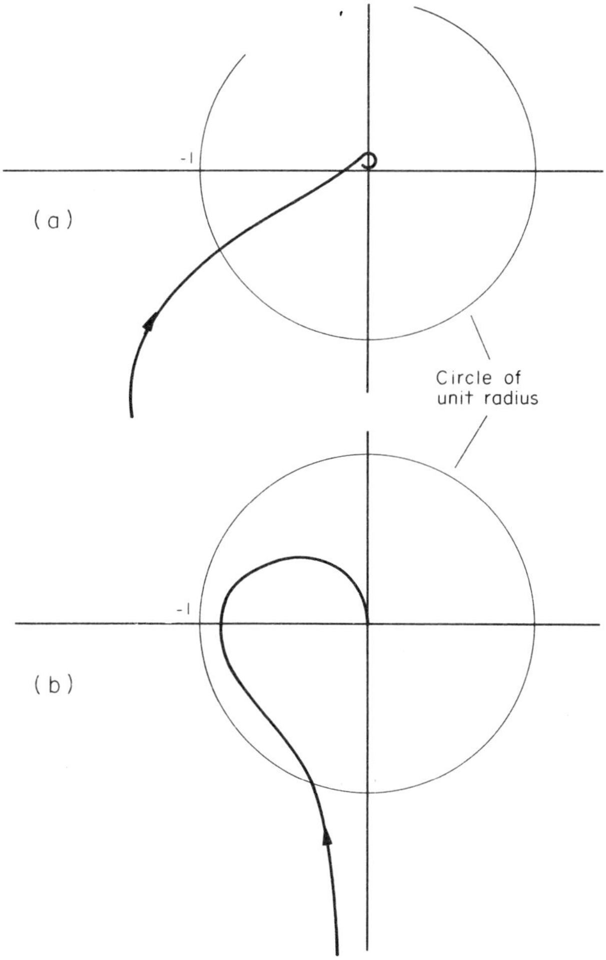

Fig. 5.46. Gain and phase margin combinations: (a) Large GM, small PM. (b) Small GM, large PM.

CHAPTER 6

ANALYSIS OF NON-LINEAR SYSTEMS

6.1 INTRODUCTION TO NON-LINEAR CONTROL

The mathematical description of a dynamic system can be embodied in a differential equation such as equation 5.11 (repeated here for reference):

$$a\,\ddot{\theta}(t) + b\,\dot{\theta}(t) + c\,\theta(t) = f(t) \qquad (5.11)$$

and such a system would be described as linear if the coefficients a, b and c are not affected by the values of the dependent variable $\theta(t)$ or of the independent variable f(t). For linear systems we have available all the relatively straightforward methods of analysis and prediction which were described in Chapter 5, based on the Laplace Transform.

In fact we can recognise that even in a simple case such as the heat-exchanger example in Section 5.4, the assumption that the coefficients are constant is not strictly true. However, provided the range of temperature involved is not too great then the heat-transfer coefficients will be nearly constant and thermal properties will not change too much, so the assumption of linear behaviour is justified in that it permits the use of Laplace Transform techniques for analysis. The analysis is then generally systematic and straightforward. The type of controller to which this method is appropriate is one with a continuously variable output, applied to modulate a final control element, such as a valve or damper. This type of controller and control element is relatively complicated in design and expensive in first cost. A type of controller which is much cheaper and simpler is one which has only a limited number of discrete output states, usually only two. The methods of analysis outlined in the previous chapter are not generally applicable to this type of relay controller or switching controller. Where the controller output can take on one of only two values the characteristic is usually termed ON/OFF control. Where the controlled variable is temperature, the characteristic is also termed *thermostatic* control, since most room thermostats are constructed to operate as switches. The action of the switch, in opening on a rise in temperature and closing on a fall in temperature, introduces changes in the mathematical description of the system dynamics. A linear dynamic system is linear because the equations describing its behaviour have constant coefficients, which are invariant with time or with the value of the dependent or independent variables. If this condition is not met then the system is non-linear. Clearly there are many more systems conceivable which do not satisfy the condition for linearity than there are systems which do satisfy it. Also, the non-linearity may be of very many different types. Hence, one cannot expect to find methods which are generally applicable to *any* non-linear system but only to those systems with particular types of non-linearity.

We are concerned here with relay control and there have been three methods published for non-linear system analysis which are suitable for relay systems. They are known as the Step Response method, The Describing Function method, and Tsypkin's method. Each method will be described in detail in the following sections.

Each method assumes a system in which the linear and non-linear components can be separated, as shown in Fig. 6.1, where $L(s)$ is the Laplace transfer function of the

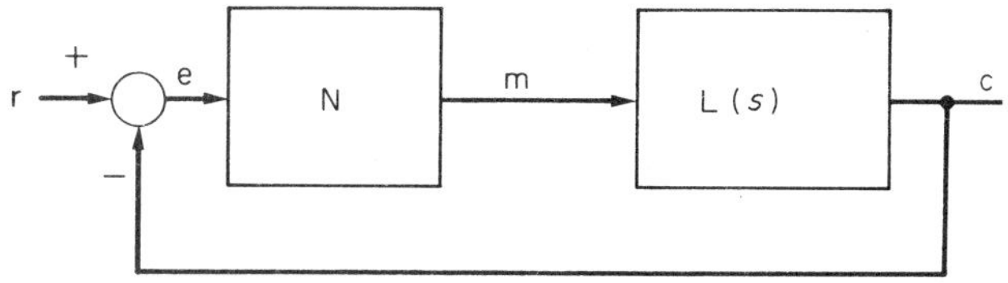

Fig. 6.1. Block diagram of general non-linear system: non-linear element N, linear element $L(s)$.

linear part of the system and N represents the non-linear part. Generally, N is the input/output characteristic of the relay or thermostat. The types of characteristic to be considered are shown in Fig. 6.2. The output of N can normally

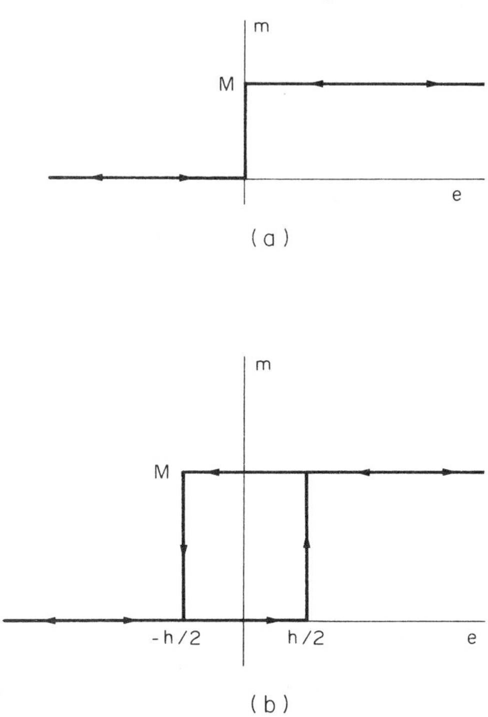

Fig. 6.2. Relay switching characteristics. (a) Ideal relay. (b) Relay with hysteresis.

take on one of only two values: zero or +M (corresponding to a zero rate of heat supply or a constant positive rate of heat supply). For the ideal relay, as the error signal e changes its sign, the output m immediately switches over to its alternative value. For the relay with hysteresis, the magnitude of the error signal has to exceed zero by the amount $h/2$ before switching will occur in either direction. Most commercial room thermostats can be represented by the characteristic shown in Fig. 6.2(b) with h equal to about 2°C. The value of h is usually fixed by the manufacturer by the choice of dimensions, spring strength, etc., although it can be varied by the user in some designs.

6.2 THE STEP RESPONSE METHOD

Considering the general system shown in Fig. 6.1, with N representing either of the characteristics shown in Fig. 6.2, when continuous steady oscillations exist, the output of N will be a square wave, a train of rectangular pulses, of amplitude M, as shown in Fig. 6.3. This wave train can be considered to be the algebraic

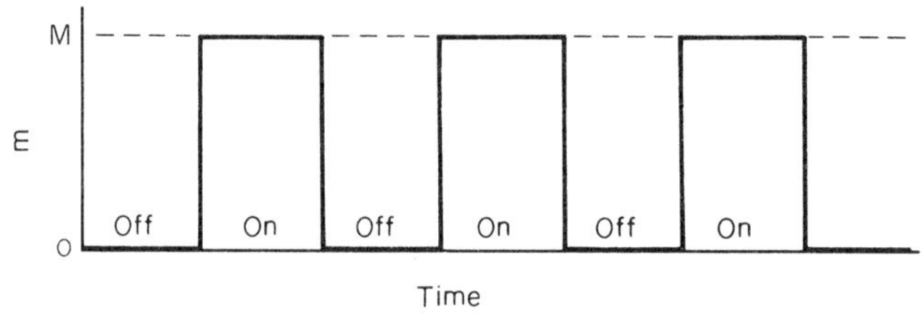

Fig. 6.3. Output wave of non-linear switching element.

sum of a series of positive and negative delayed step functions, all of magnitude M, as shown in Fig. 6.4. This wave train is the input signal m(t) to the linear part of the system $L(s)$. Since $L(s)$ is linear then the principle of superposition must apply and its output response to the input m(t) must be the algebraic sum of its responses to the step function inputs. Thus, if the response of $L(s)$ to a unit step function is known, say q(t), then the response to the wave of Fig. 6.3 can be found by addition, either graphically or numerically. Figure 6.5 illustrates the complete procedure, taking a controller with hysteresis h.

To fix ideas, the application to a heated room will be considered. The system is assumed to be switched on from cold, with a reference value (set point) on the thermostat of r°C. If the time response of the room temperature to a heat input step of *unit* magnitude is q(t), then the response to a step of magnitude M will be Mq(t), i.e. M times as large. The room temperature will rise initially according to the function Mq(t) until it exceeds the upper switching point $r+h/2$. At that instant ($t=t_1$) a switching will occur and the heat supply will be cut off. This will initiate a contribution of $-Mq(t-t_1)$ to the response curve, which is now the function

$$Mq(t) - Mq(t-t_1) \qquad t_1 < t < t_2.$$

This eventually starts to come down and reduces below the lower switching point $r-h/2$. At this point ($t=t_2$) another switching will occur, initiating a further addition term to the response, which now has the function

$$Mq(t) - Mq(t-t_1) + Mq(t-t_2) \qquad t_2 < t < t_3.$$

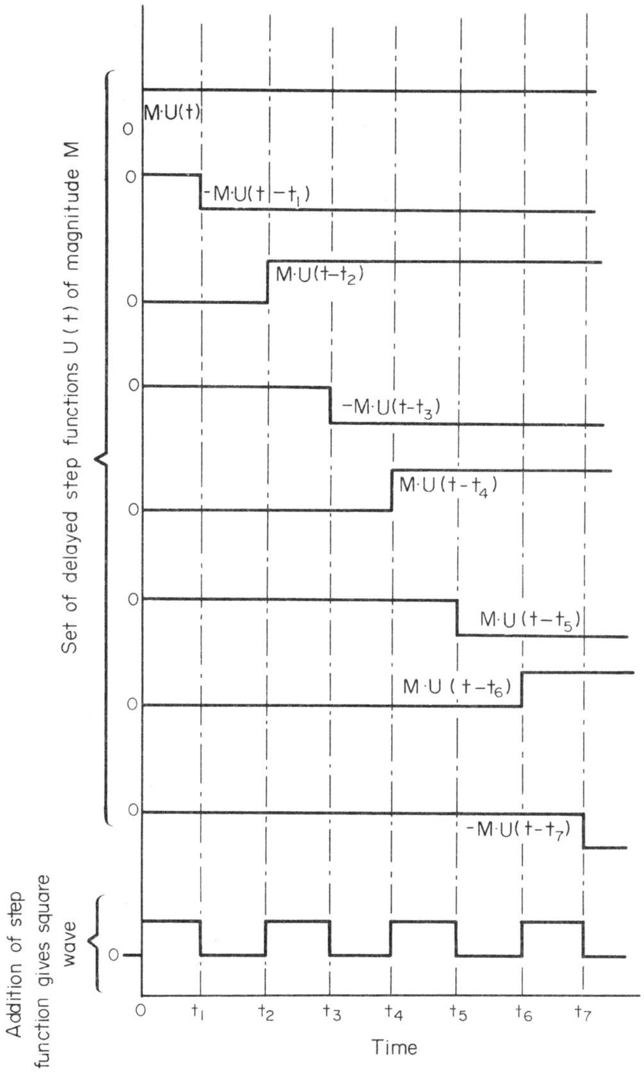

Fig. 6.4. A square wave as the sum of a set of step functions.

In this way the time history of the response is built up, and after a number of cycles the controlled temperature settles down to steady oscillations of constant amplitude and period. At any instant between the n-th and (n+1) th switching points, the response c(t) will be given by a suitable summation of delayed and inverted response curves.

$$c(t) = M \sum_{i=0}^{n} (-1)^i q(t-t_i), \quad t_n < t < t_{n-1}. \tag{6.1}$$

This method allows the calculation of the value of the controlled variable at any instant. In order to find the amplitude, period and mean value of the controlled oscillations it would be necessary to evaluate c(t) for a large number of points on

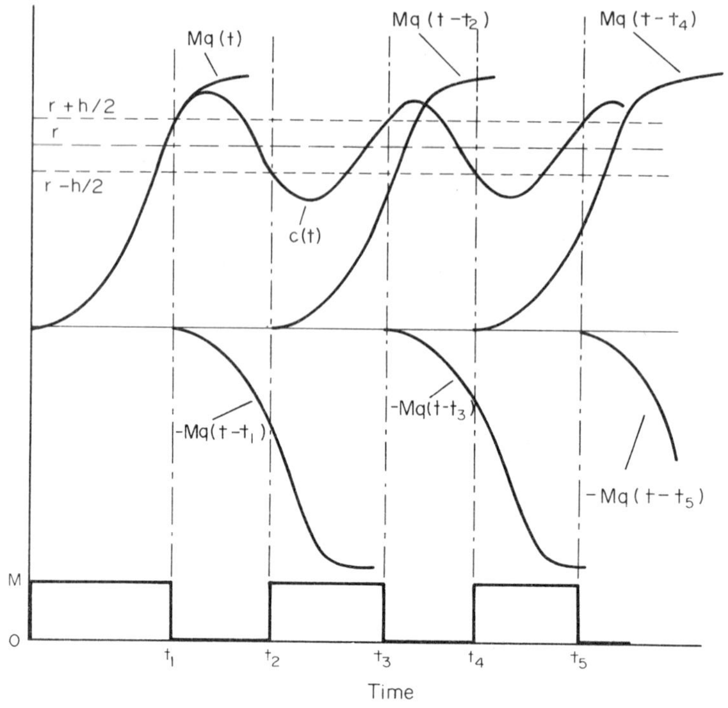

Fig. 6.5. The build-up of the controlled oscillation.

the cycle: the method is numerical, not analytical.

In principle this method is completely flexible. It allows any non-linear characteristic to be used, including dead zone, multiple step output and time-varying parameters. It is an exact method in that no approximating assumptions are made, and the value of the response at any time can be found with as much accuracy as is justified by the precision with which the original unit step response q(t) is known.

Because of the step-by-step nature of the calculation the method has some limitations:

1. Studying the influence of the system parameters on the controlled response will involve considerable computational labour, since the whole calculation has to be repeated each time any parameter is changed. Obviously, a digital computer program is the most convenient way of performing the calculations.

2. The early behaviour of the system soon after switch-on can be found quite easily, but a rather large number of cycles may have to be followed before a steady cyclic state is reached.

EXAMPLE 6.1. The step response of a room to a heat input was tested by switching on from cold a 1-kW heater. The room temperature was recorded until it reached a steady state. The room-temperature response could be accurately modelled by the function

$$q(t) = 10\ (1 - (1 + t)\ \exp\ (-t))$$

where t is the time after switch-on (minutes) and q(t) is the rise above outside temperature.

Use the step-response method to predict the period and amplitude of controlled oscillations when the room temperature is controlled by a switching thermostat which has a total hysteresis of 2°C. The thermostat set point settings should be taken as (a) 23°C, (b) 16°C and (c) 5°C to illustrate the performance over a range of loads. The room heater used is of 2.5 kW power and is of the same type as that used for the test above. The outside temperature may be assumed to remain constant at zero.

Since the step response test was done using a 1-kW heater, and the thermostat is to be used with one of 2.5 kW power, the value of M in this case is 2.5, and the function Mq(t) will be

$$Mq(t) = 25\ (1 - (1 + t)\exp(-t)).$$

Even with a relatively simple function such as this one, the manual evaluation and summation of the contributing terms rapidly becomes very tedious and mistake-prone. It is much more convenient and accurate to use a digital computer. The listing of a suitable FORTRAN program is given below. The controlled responses calculated using this program are shown in Fig. 6.6, where the on/off switching cycles are also illustrated. Note the change in cycle shape and overshoot as the load changes.

Note also that the ratio: ON time/Periodic time is equal to the ratio: Mean temperature rise/Maximum possible steady temperature rise (in this case 25.0°C). The results are summarised as follows:

Set point (°C)	ON time (min)	OFF time (min)	Periodic time (min)	Peak-to-peak amplitude (°C)	Mean temp. rise (°C)
23.0	3.70	0.51	4.21	4.0	21.98
16.0	1.39	0.83	2.22	3.3	15.61
12.5	1.06	1.06	2.12	3.2	12.50
5.0	0.64	2.09	2.73	3.5	5.84

6.3 THE DESCRIBING FUNCTION METHOD

This method can be considered as an attempt to extend the concept of the linear transfer function to use in non-linear systems. For a linear transfer function with a sinusoidal input of unit amplitude sin (ωt) the output will be a sine wave $A(\omega)\sin(\omega t + \phi(\omega))$. The transfer function introduces an amplification $A(\omega)$ and phase shift $\phi(\omega)$ both of which may in general be dependent on frequency ω but not (for a linear system) on the amplitude. The difficulty in extending this concept to non-linear systems is that the output waveform for such systems is not in general sinusoidal, and in certain cases may not be of the same frequency as the output wave. It is also in many cases amplitude-dependent.

If the non-sinusoidal periodic output waveform of a non-linear element is subjected to Fourier analysis, it is broken down into components of particular frequencies,

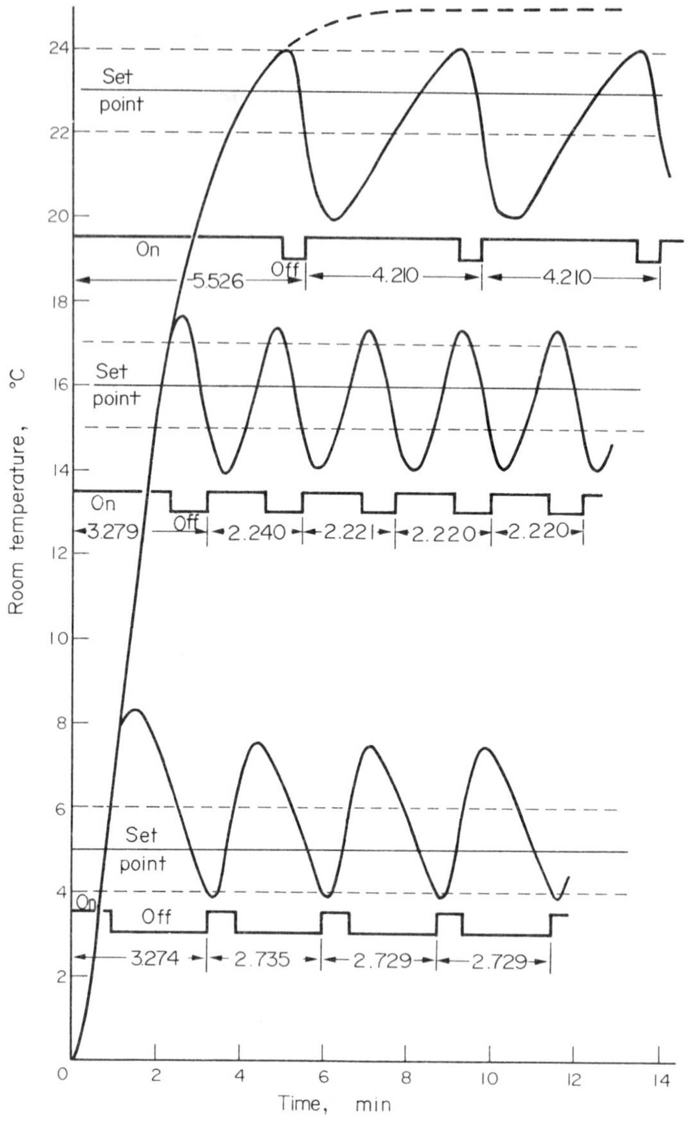

Fig. 6.6. Results of Example 6.1.

called harmonics because their frequencies are in the ratios of the natural numbers, the harmonic series. An *approximate* equivalent transfer function $N(X)\phi(X)$ can then be defined as

$$N(X) = \frac{\text{Amplitude of the fundamental component of the ouput (m)}}{\text{Amplitude (X) of the sinusoidal input (e)}},$$

$\phi(X)$ = Phase shift of the fundamental component of the output with reference to the sinusoidal input.

```
      PROGRAM THSTAT (INPUT,OUTPUT,TAPE1=INPUT,TAPE2=OUTPUT)
C THIS PROGRAM USES THE STEP RESPONSE METHOD TO CALCULATE THE
C CONTROLLED CYCLING CHARACTERISTICS OF A SYSTEM WITH A SWITCHING
C CONTROLLER OF SET POINT'S' AND TOTAL HYSTERESIS BAND WIDTH 'H' .
C THE STEP RESPONSE FUNCTION IS DEFINED IN THE SUBPROGRAM F(X).
C THE FUNCTIONS ARE SUMMED IN THE SUBPROGRAM RESP(Y,M).
C
      DIMENSION T(100),P(100),B(100)
      COMMON T
  200 FORMAT(1H1)
      AMAX=F(1.E6)
      SETTLE=1.E-6*AMAX
C READ IN THE VALUES OF SET POINT AND HYSTERESIS
    7 READ(1,100) S,H
      IF(S.LE.0.) STOP
      WRITE(2,200)
  100 FORMAT(2F8.2)
      WRITE(2,208)
  208 FORMAT(1H ,100(1H*))
      WRITE(2,201) S,H
  201 FORMAT(1H0,4X,'SET POINT ',F8.2,5X,'HYSTERESIS ',F8.2)
C TEST FOR THE EXISTENCE OF A CYCLE
      IF((S-0.5*H).LE.0.) GOTO 40
      IF((S+0.5*H).GE.AMAX) GOTO 50
      WRITE(2,205)
  205 FORMAT(1H0,13X,'TIME', 9X,'RESPONSE',/)
C SET TIME TO ZERO
      TI=0.
      N=1
      T(1)=0.
C SET INITIAL STEP SIZE TO 10.
    2 A=10.
      SW=S-(0.5*H)*FLOAT((-1)**N)
    3 Q=RESP(TI,N)
      L=N+1
      I=(-1)**L
C CHECK IF RESPONSE HAS EXCEEDED SWITCHING LEVEL
      IF(I*Q.LT.I*SW) GOTO 4
C CHECK IF RESPONSE IS CLOSE ENOUGH TO SWITCHING LEVEL
      IF(ABS(Q-SW).GT.1.E-8) GOTO55
      GOTO 5
C INCREMENT THE TIME BY ONE STEP
    4 TI=TI+A
      IF(TI.GT.1000.) STOP
      GOTO3
   55 TI=TI-A
C DIVIDE STEP SIZE BY TEN
      A=0.1*A
      GOTO3
    5 N=N+1
      T(N)=TI
C PRINT TEN EQUAL-SPACED POINTS ON THE CYCLE
      T2=0.1*(T(N)-T(N-1))
      DO 6 K=1,10
      T3=T(N-1)+K*T2
      R=RESP(T3,(N-1))
    6 WRITE(2,203) T3,R
```

```
  203 FORMAT(1H ,11X,F8.3, 3X,F10.3)
      SUM=0.
      M=0
      IF(N.LT.3) GOTO 2
      T4=0.01*(T(N)-T(N-1))
      DO 9 K1=1,100
      T5=T(N-1) + K1*T4
      SUM=SUM+RESP(T5,N-1)
    9 M=M+1
      AMEAN=SUM/M
      B(N)=AMEAN
       AN=0.5*FLOAT(N)
C IF N IS EVEN, CALCULATE THE PERIOD
      IF(((AN-AINT(AN)).LT.0.2))GOTO 2
      TON=T(N-1)-T(N-2)
      TOFF=T(N)-T(N-1)
      P(N-1) = T(N) - T(N-2)
      IF(N.LT.4) GOTO2
      AVER=((B(N)*TOFF+B(N-1)*TON)/P(N-1))
      WRITE(2,209)TON,TOFF ,P(N-1)
  209 FORMAT(1H ,10X,'ON TIME= ',F8.4,4X,'OFF TIME= ',F8.4,
     14X,'PERIOD= ',F8.4)
      WRITE(2,204) AVER
  204 FORMAT(1H ,'      MEAN RESPONSE OVER ONE CYCLE= ',F10.3)
      IF((ABS(F(TI)-AMAX)).GT.SETTLE) GOTO2
      WRITE(2,208)
      GOTO 7
   40 WRITE (2,41)
   41 FORMAT(1H ,'CYCLING WILL NOT OCCUR, SET PT TOO LOW')
      WRITE(2,208)
      GOTO 7
   50 WRITE(2,51)
   51 FORMAT(1H ,' CYCLING WILL NOT OCCUR, SET PT TOO HIGH')
      WRITE(2,208)
      GOTO 7
      END
      FUNCTION RESP(Y,M)
      DIMENSION T(100)
      COMMON T
      B=0.
      DO 8 J=1,M
    8 B=B-((-1)**J)*F(Y-T(J))
      RESP=B
      RETURN
      END
      FUNCTION F(X)
      IF (X.GT.200.) X=200.
      F=25.*(1.-EXP(-X))
      RETURN
      END
```

There are two important differences with respect to the linear case:

1. There is an error introduced by the approximation of replacing the output waveform by its fundamental Fourier component.

2. The resulting amplification factor and phase shift will in general be dependent both on the frequency and on the amplitude of the input sine wave. In fact we are concerned here with non-linearities which are solely amplitude dependent, and not frequency dependent. The former non-linearities are the ones most commonly found.

The describing function method in fact generates two loci in the complex plane. One is produced by the variation with frequency of the response of the linear part of the system, $L(s)$; the other produced by the variation with amplitude of the equivalent sinusoidal transfer function of the non-linear part, N. An intersection of these two loci indicates that continuous oscillations of given amplitude and frequency are predicted.

It was shown in Section 5.7 that the stability of a controlled system depends upon the sign of the real parts of the roots of its characteristic equation. Any root with a positive real part will produce a non-decaying term in the time response and hence cause instability. In particular, the case with roots which are purely imaginary has a time response which is a steady oscillation of constant amplitude. This case corresponds to the steady cycling response of the system under thermostatic control, with oscillations which neither build up nor die away.

If we apply this to the non-linear case, the corresponding characteristic equation is (compare Section 5.7):

$$1 + N \cdot L(s) = 0 \qquad (\text{where } N = N(X)\phi(X))$$

and we seek the solution for $s = j\omega$, i.e. purely imaginary roots. The equation above becomes

$$1 + N \cdot L\,(j\omega) = 0,$$

hence

$$-1/N = L(j\omega),$$

so that an intersection of the locus of the complex quantity $-1/N(X)\phi(X)$ with the locus of $L(j\omega)$ implies that a steady oscillation can exist with amplitude X and frequency ω. The linear part of the system determines the locus $L(j\omega)$ as frequency ω increases from zero to infinity. The non-linear part determines the locus $-1/N(X)\phi(X)$ as the amplitude X increases from zero to infinity. The locus $-1/N(X)\phi(X)$ is called the describing function.

We shall obtain the describing functions for a number of non-linearities of interest.

6.3.1 Ideal Relay

For an ideal relay (that is, one with zero hysteresis) the output of the non-linear element will be a square wave of amplitude M, as shown in Fig. 6.7. The changeovers of the output wave between zero and +M will occur as the input sine wave crosses through zero. Fourier analysis of the square wave of peak-to-peak amplitude M and equal on and off times gives the series

$$m(t) = (2M/\pi) \sum_{n=1}^{\infty} (1/n) \sin n\omega t \qquad (n \text{ odd}), \tag{6.2}$$

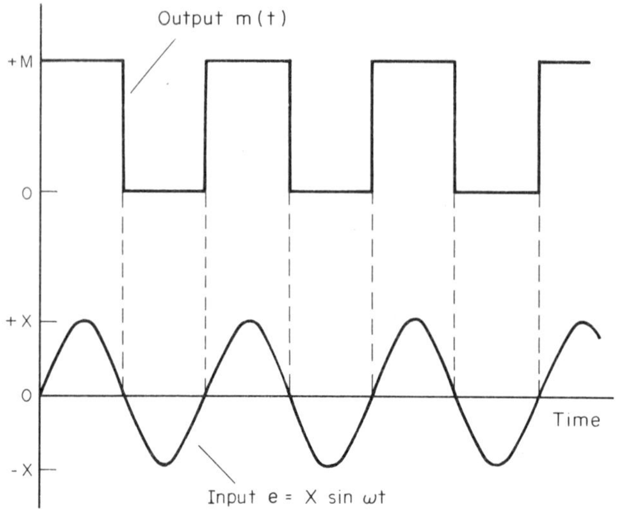

Fig. 6.7. Input and output waves for ideal relay.

and the describing function is given by the amplitude of the first term of the series, $2M/\pi$, divided by the amplitude of the input sine wave, X. The phase shift is zero:

$$N(X) = 2M/\pi X, \ \phi(X) = 0,$$

hence

$$-1/N = -\pi X/2M, \tag{6.3}$$

which is a negative real number, lying on the negative half of the real axis and running from the origin when $X = 0$ to negative infinity when $X = \infty$. This is shown in Fig. 6.8 for the case when $h = 0$.

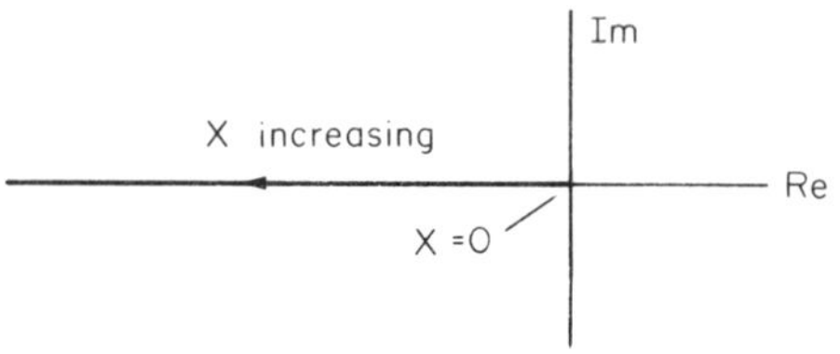

Fig. 6.8. Describing function locus for ideal relay ($h = 0$).

6.3.2 Relay with Hysteresis h

For the case where the relay has a total hysteresis h then the input signal $e = X \sin \omega t$ has to reach the value $h/2$ before switching of the output takes place. There is thus a phase lag ϕ between input and output given by

$$e = X \sin \phi = h/2$$

which gives

$$\phi = \arcsin h/2X. \tag{6.4}$$

If the input amplitude X is less than $h/2$, no switching occurs and the output amplitude is zero. If the input amplitude is greater than or equal to $h/2$ then the output is a square wave of amplitude M and with a phase lag of arcsin ($h/2X$), as shown in Fig. 6.9.

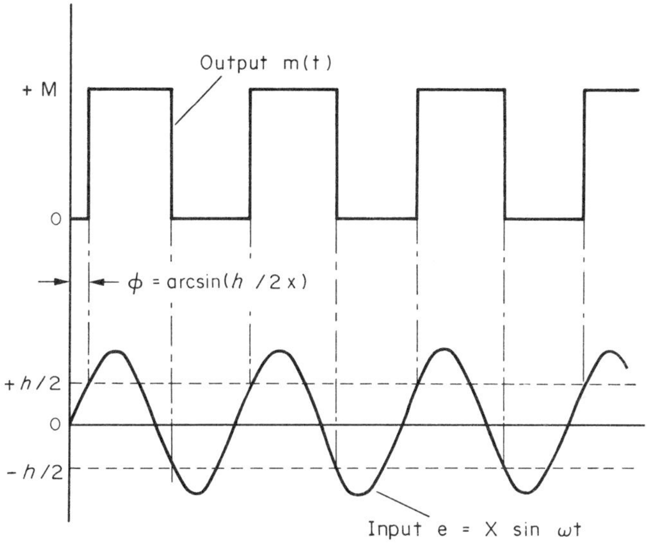

Fig. 6.9. Input and output waves for relay with hysteresis h.

Hence

$$N = \begin{cases} 0 \text{ for } X < h/2 \\ \dfrac{2M}{\pi X} \underline{/\arcsin (h/2X)} \text{ for } X \geq h/2 \end{cases}$$

and

$$-1/N = \frac{\pi X}{2M} \underline{/\pi + \arcsin (h/2X)} \quad X \geq h/2$$

and in cartesian form:

$$-1/N = \frac{-\pi}{4M}\left[4X^2 - h^2\right]^{\frac{1}{2}} - j\,\frac{\pi h}{4M} \tag{6.5}$$

which shows that the imaginary part is independent of X, giving a locus which is a straight line parallel to the real axis, as in Fig. 6.10, for $h/2 < X < \infty$.

EXAMPLE 6.2. Use the describing function method to predict the amplitude and period of oscillations when a linear system whose transfer function is

$$F(s) = 25/(1 + 2000s)^2 \qquad {}^{o}C/kW$$

is controlled by a switching thermostat whose total hysteresis is $3^{o}C$. The value of M may be taken as unity, corresponding to a heater power of 1 kW.

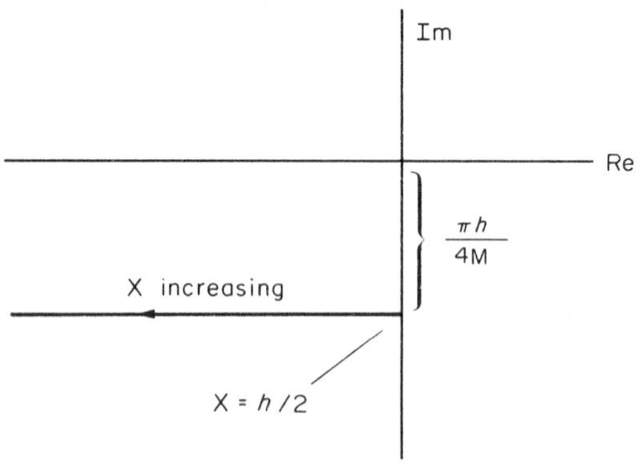

Fig. 6.10. Describing function locus for relay with hysteresis h.

The describing function locus will be a straight line parallel to the negative real axis and displaced from it by the distance $-\pi h/4M$, as in Fig. 6.10, so that, from equation (6.5):

$$-1/N = -0.785\ (4X^2 - 9)^{\frac{1}{2}} - j2.36$$

which is a function of X as X varies from 1.5 to $+\infty$.

The frequency response locus of the linear part can be obtained by substituting $j\omega$ for s in the transfer function given above, as shown in Section 5.10. Substituting and rationalising gives:

$$F(j\omega) = (25-10^8\omega^2)/C.D. - j\omega 10^5/C.D.$$

where the common denominator C.D. is

$$(1-4\times10^6\ \omega^2)^2 + 1.6\times10^7\ \omega^2.$$

These loci are shown in Fig. 6.11, and the intersection occurs at the point -2.49-j2.36, at an angular frequency $\omega = 0.00126$ radian/second.

The amplitude X is given by

$$-2.49 = -0.785\ (4X^2-9)^{\frac{1}{2}}$$

from which X = 2.18°C.

The period T is given by $2\pi/\omega$.

$$T = 2\pi/0.00126 = 5000 \text{ seconds.}$$

The switching period will be over an hour.

6.4 TSYPKIN'S METHOD

This method represents an attempt to overcome the inaccuracies inherent in the describing function method, by including the higher harmonics of the Fourier series. In principle it can give exact results for the frequency of oscillation,

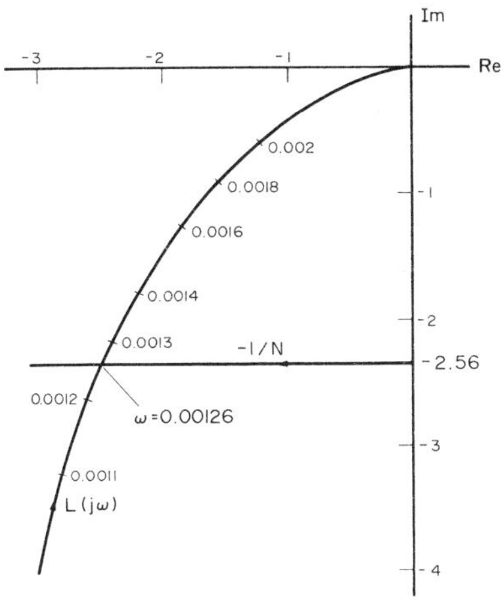

Fig. 6.11. Describing function solution for Example 6.2.

but exact values of the amplitude are rather difficult to obtain. An approximate result for the amplitude is available.

6.4.1 Switching Conditions

We introduce the analysis by considering an arbitrary steady oscillation in the error signal $e(t)$ applied to the input of the non-linear element N. Taking zero time as the instant at which $e(t)$ first exceeds $h/2$, a switching will occur at that instant, as shown in Fig. 6.12. If the oscillation is symmetrical in the

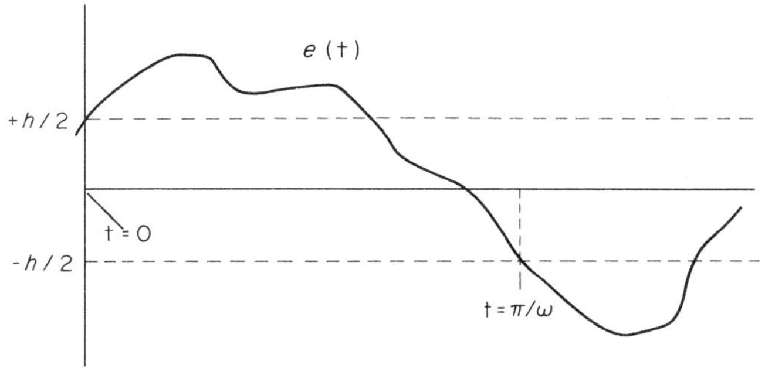

Fig. 6.12. Arbitrary periodic oscillation of $e(t)$.

sense that a switching in the opposite direction occurs half a time period later, then we can write general conditions for the existence of a cycle, as follows:

$$\left.\begin{aligned} & e(\pi/\omega) = -h/2 \\ \text{and} \quad & \frac{d}{dt}\,(e(\pi/\omega)) < 0. \end{aligned}\right\} \qquad (6.6)$$

For an ideal relay, with $h = 0$, the corresponding conditions are

$$\left.\begin{aligned} e(\pi/\omega) &= 0 \\ \text{and} \quad \frac{d}{dt}\,(e(\pi/\omega)) &< 0. \end{aligned}\right\} \qquad (6.7)$$

Alternative switching conditions can be written for other relay characteristics.

6.4.2 Analysis of Steady Cycling

If switchings occur at times 0, π/ω, $2\pi/\omega$, $3\pi/\omega$, etc., then the output of the non-linear element will be a square wave of equal on and off times and peak-to-peak amplitude M. As mentioned for the describing function method, the Fourier series for such a wave is

$$m(t) = 2M/\pi \sum_{n=1}^{\infty} (1/n) \sin n\,\omega t \qquad n = 1, 3, 5, 7...$$

Considering $L(j\omega)$, the sinusoidal response of the linear transfer function $L(s)$, we can express this in complex form:

$$L(j\omega) = A(\omega) \exp\,(j\phi(\omega))$$

where

$$A(\omega) = \left[(\text{Re } L\,(j\omega))^2 + (\text{Im } L(j\omega))^2\right]^{\frac{1}{2}}$$

and $\phi(\omega) = \arctan\,[\text{Im } L\,(j\omega)/\text{Re } L\,(j\omega)]$.

The response of the linear element $L(s)$ to a sine wave input of frequency ω and unit amplitude is given by

$$A(\omega) \sin\,(\omega t + \phi(\omega)).$$

Hence, the response of the linear element to the series of sine waves which make up the Fourier analysis of the square wave will be

$$c(t) = 2M/\pi \sum_{n \text{ odd}} (1/n)\,A(n\omega) \sin\,(n\omega t + \phi(n\omega)) \qquad (6.8)$$

and the time derivative of the response is

$$\dot{c}(t) = 2M\omega/\pi \sum_{n \text{ odd}} A(n\omega) \cos\,(n\omega t + \phi(n\omega)). \qquad (6.9)$$

For the regulator case, where it is desired to control at a constant condition, we can measure all variables from the reference input r as zero. We can thus take the error variable e to be the negative of the controlled variable c so that $e = -c$ and the switching conditions of equation (6.6) can be combined with equations (6.8) and (6.9) to give

$$\begin{aligned} e(\pi/\omega) &= -\,c(\pi/\omega) \\ &= 2M/\pi \sum_{n \text{ odd}} (1/n)\,A(n\omega) \sin \phi(n\omega) \\ &= -h/2 \text{ from equation (6.6)} \qquad (6.10) \end{aligned}$$

(since $\sin(\pi+\phi) = -\sin\phi$)

$$\dot{e}(\pi/\omega) = -\dot{c}(\pi/\omega)$$

$$= 2M\,\omega/\pi \sum_{n\ \mathrm{odd}} A(n\omega)\cos\phi(\omega)$$

$$< 0 \text{ from equation (6.6)} \qquad (6.11)$$

(since $\cos(\pi+\phi) = -\cos\phi$)

Now we introduce a function $T(\omega)$ whose real and imaginary parts are constructed from the functions in equations (6.10) and (6.11):

$$T(\omega) = 1/\omega\ \dot{e}(\pi/\omega) + j\ e(\pi/\omega) \qquad (6.12)$$

so that the switching conditions of equations (6.6) and (6.7) become

$$\mathrm{Im}\ T(\omega) = -h/2 \text{ and } \mathrm{Re}\ T(\omega) < 0 \qquad (6.13)$$

by comparison with equations (6.10) and (6.11) respectively.

We can now define the Tsypkin locus $T(\omega)$ by analogy with the Nyquist locus. It is the path of the complex number $T(\omega)$ in the complex plane as ω increases from 0 to infinity. From equation (6.13) it appears that the periodicity conditions in equation (6.6) will be satisfied if the locus $T(\omega)$ intersects the horizontal line $-h/2$ in the left-hand plane. If this occurs at the point $T(\omega_1)$ an oscillation at a frequency ω_1 is indicated, as shown in Fig. 6.13.

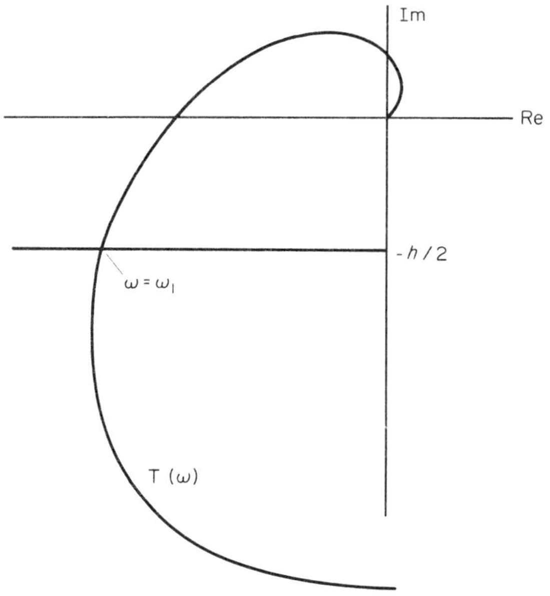

Fig. 6.13. Intersection of Tsypkin's locus $T(\omega)$ with line $-h/2$ in the left-hand plane.

6.4.3 Constructing the Tsypkin Locus T(ω) from the Nyquist Locus L(jω)

If the frequency response of $L(s)$, the linear part of the system shown in Fig. 6.1, has real and imaginary parts:

$$L(j\omega) = U(\omega) + j\,V(\omega) \tag{6.14}$$

then we can write

and

$$\left.\begin{aligned} U(\omega) &= A(\omega)\cos\phi(\omega) \\ V(\omega) &= A(\omega)\sin\phi(\omega) \end{aligned}\right\} \tag{6.15}$$

as indicated in Fig. 6.14, then the real part of $T(\omega)$ is seen from equations (6.11)

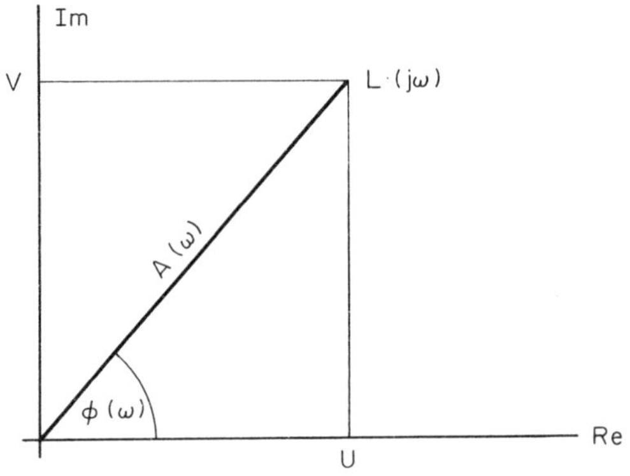

Fig. 6.14. Real and imaginary parts of $L(j\omega)$.

and (6.12) to be given by

$$\begin{aligned} \text{Re } T(\omega) &= \frac{2M}{\pi} \sum_{n \text{ odd}} A(n\omega)\cos\phi(n\omega) \\ &= \frac{2M}{\pi} \sum_{n \text{ odd}} U(n\omega) \end{aligned} \tag{6.16}$$

and the imaginary part of $T(\omega)$ is given by

$$\begin{aligned} \text{Im } T(\omega) &= \frac{2M}{\pi} \sum_{n \text{ odd}} \frac{1}{n} A(n\omega)\sin\phi(n\omega) \\ &= \frac{2M}{\pi} \sum_{n \text{ odd}} \frac{1}{n} V(n\omega) \end{aligned} \tag{6.17}$$

from equations (6.10) and (6.12).

Points on the $T(\omega)$ locus can be evaluated by a digital computer program from the expressions in equations (6.16) and (6.17) or $T(\omega)$ can be constructed graphically from the $L(j\omega)$ locus, as shown schematically in Fig. 6.15. At high frequencies $T(\omega)$ coincides with $L(j\omega)$.

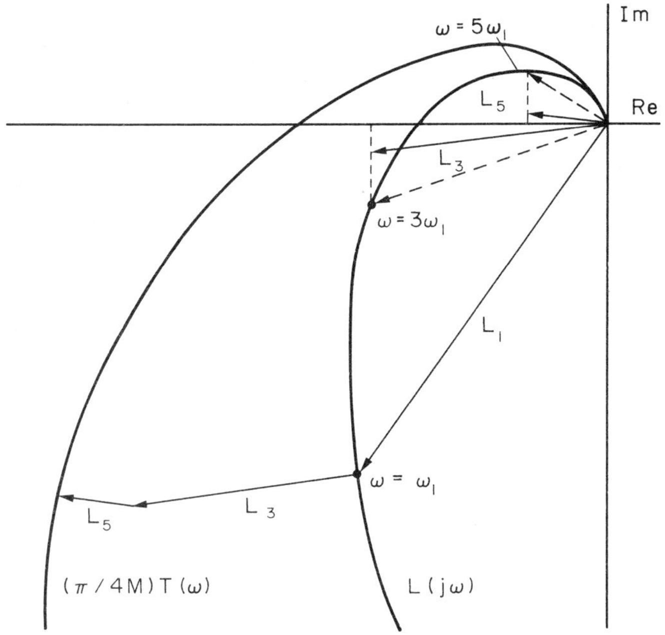

Fig. 6.15. Graphical construction of the Tsypkin locus from the Nyquist locus. Note the multiplying factor $2M/\pi$ is to be included.

6.4.4 Amplitude of the Oscillation

The method described above gives an explicit result for the frequency of oscillation ω_1. The amplitude of oscillation is not given explicitly but an approximate value can be computed in two ways.

Referring to equation (6.9) and taking only the first term of the summation (n=1), the derivative $\dot{c}(t)$ will be zero at the maximum and minimum values of the waveform:

$$\dot{c}(t) = \frac{2M\omega_1}{\pi} A(\omega_1) \cos(\omega_1 t + \phi(\omega_1)) = 0$$

where the frequency ω_1 is that given by the analysis of Section 6.4.2. For this condition to be satisfied we require

$$\cos(\omega_1 t' + \phi(\omega_1)) = 0$$

where t' denotes the time of occurrence of the maximum value of c(t).

Hence $\omega_1 t' + \phi(\omega_1) = \pi/2$

and $t' = (\pi/2 - \phi(\omega_1))/\omega_1$.

Similarly for the time t" at which the minimum value of c(t) occurs we can obtain

$$t'' = (3\pi/2 - \phi(\omega_1))/\omega_1$$

and substituting these values of time into equation (6.8) gives

$$c(t') = c_{max}$$

$$\simeq (2M/\pi)\ A(\omega_1)\ \sin\ (\omega_1 t' + \phi(\omega_1))$$

$$= (2M/\pi)\ A(\omega_1)\ \sin\ (\pi/2)$$

$$= 2MA\ (\omega_1)/\pi$$

and

$$c(t'') = c_{min}$$

$$\simeq (2M/\pi)\ A(\omega_1)\ \sin(\omega_1 t'' + \phi(\omega_1))$$

$$= (2M/\pi)\ A(\omega_1)\ \sin(3\pi/2)$$

$$= -2MA(\omega_1)/\pi$$

so that an approximate value of the peak-to-peak amplitude will be given by

$$c(t') - c(t'') = 4MA(\omega_1)/\pi. \quad (6.18)$$

A better approximation to the peak-to-peak amplitude can be obtained by numerical evaluation of c(t) from equation (6.8) for $\omega = \omega_1$ and for a large number of closely spaced values of t in the range zero to $2\pi/\omega_1$, in this case taking the summation into account to as many terms as necessary. The amplitude can then be obtained by subtracting the smallest value of c(t) from the largest value.

EXAMPLE 6.3. A process whose linear transfer function is $L(s)$ is controlled by a relay with a total hysteresis of 2°C, operating a heater of 3 kW output. If $L(s) = 6/(1+s)^3$ °C/kW, find the period and amplitude of the controlled oscillations.

The Nyquist locus $L(j\omega)$ is obtained by substituting $j\omega$ for s and rationalising:

$$L(j\omega) = 6/(1 + j\omega)^3$$

$$= 6(1 - 3\omega^2)/CD + j6\omega(\omega^2 - 3)/CD$$

where

$$CD = (1 - 3\omega^2)^2 + \omega^2(3 - \omega^2)^2.$$

Hence, from the real and imaginary parts of $L(j\omega)$, we have the following functions for $U(\omega)$ and $V(\omega)$:

$$U(\omega) = 6(1-3\omega^2)/((1-3\omega^2)^2 + \omega^2(3-\omega^2)^2)$$

$$V(\omega) = 6\omega(\omega^2-3)/((1-3\omega^2)^2 + \omega^2(3-\omega^2)^2)$$

from which

$$A(\omega) = 6/((1-3\omega^2)^2 + \omega^2(3-\omega^2)^2)^{\frac{1}{2}}$$

and

$$\phi(\omega) = \arctan\ (\omega(\omega^2-3)/(1-3\omega^2))$$

and since in this case M = 3 and h = 2, we have the switching conditions

$$\text{Re } T(\omega) = 6/\pi \sum_{n \text{ odd}} U(n\omega) < 0$$

and

$$\text{Im } T(\omega) = 6/\pi \sum_{n \text{ odd}} (1/n)\, V(n\omega) = -1.0.$$

A small computer program is convenient for evaluating the real and imaginary parts of $T(\omega)$, taking as many terms of the series as is necessary for reasonable accuracy. The locus of $T(\omega)$ is shown in Fig. 6.16, along with $L(j\omega)$ for comparison. The imaginary part of $T(\omega)$ is seen to take on the value -1.0 at the frequency ω_1 = 1.29 radian/second. The period is thus predicted to be $2\pi/1.29$ = 4.87 seconds.

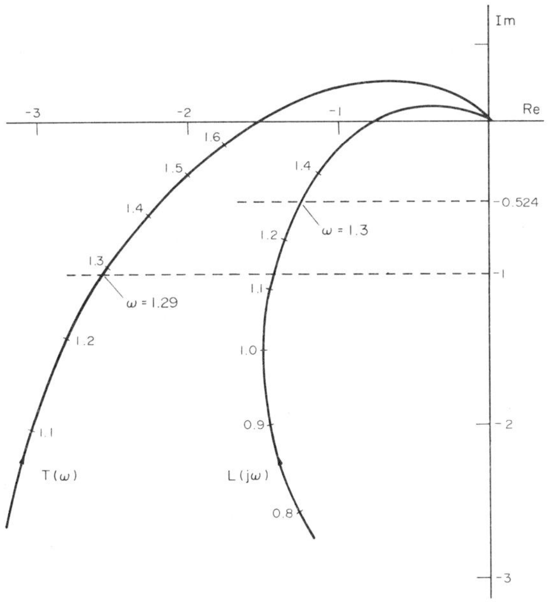

Fig. 6.16. Tsypkin locus $T(\omega)$ and Nyquist locus $L(j\omega)$ for Example 6.3.

Applying equation (6.18) with M = 3 and ω_1 = 1.29 gives 5.4°C as the approximate result for the peak-to-peak amplitude. For comparison, a computer program was written to apply the numerical procedure which is briefly described in the paragraph after equation (6.18). The period was divided into 100 equal intervals and c(t) was calculated for each of these values of time. For t = 0 and 2.45 seconds the largest values of the response c(t) were obtained, with values of -2.79

and +2.57°C respectively. The peak-to-peak amplitude is thus predicted to be 5.36°C, which matches quite closely with the approximate result given above.

Since the Nyquist locus $L(j\omega)$ for this example is also shown in Fig. 6.16, it is convenient to apply the Describing Function method to the problem as well as Tsypkin's method. Here we have $M = 3$ and $h = 2$ so $\pi h/4M = 0.524$ and we seek the intersection of $L(j\omega)$ with the horizontal line cutting the imaginary axis at a value of -0.524.

This can be seen to occur at a frequency of 1.30 rad/sec, so that the period is predicted to be $2\pi/1.30 = 4.83$ seconds. The real part of $L(j\omega)$ is then equal to -1.255, so that from equation (6.5) with $M = 3$ and $h = 2$ we have

$$-1.255 = -(\pi/12)(4X_1^2 - 4)^{\frac{1}{2}}$$

from which $X_1 = 2.6$ and the peak-to-peak amplitude is 5.2°C, which is in good agreement with the results obtained above.

6.5 COMPARISON OF THE THREE METHODS

To provide some comparison of the three methods described here, they have each been applied to a number of combinations of linear and non-linear functions. The non-linear element has the characteristic shown in Fig. 6.2, with four different values of hysteresis: zero, 1, 2 and 3°C, which lie approximately in the range of values available in commercial room thermostats. The linear function contains a gain constant of 20, with one, two or three simple lag terms in the denominator, each with a time constant of unity:

$$20/(1+s) \quad 20/(1+s)^2 \quad 20/(1+s)^3.$$

The gain constant was chosen to correspond approximately to the temperature rise above outside conditions which a heating system would be expected to maintain in a space.

The three methods of analysis already described in Sections 6.2, 6.3 and 6.4 have been applied to the various combinations of controller and process. The results are presented in Table 6.1. It can be seen from this that, where the three methods give results, they are in reasonably good agreement. For practical purposes the discrepancies are negligible. Where only the minimum data on values of the period and amplitude are required, the describing function method is to be recommended as being the least laborious. The extra accuracy of Tsypkin's method is only occasionally worth the computational effort, although there are cases where this method will give results when the describing function method will not. The step response method will always give some result. Where accurate and comprehensive information is required, the step response method is to be recommended.

TABLE 6.1 Comparison of analysis methods for non-linear systems. T = cycle period, X = cycle amplitude (peak-to-peak)

Linear transfer function	Total hysteresis	Method of analysis					
		Step response		Describing function		Tsypkin	
	h	T	X	T	X	T	X
	0	0	0	-	-	-	-
$\frac{20}{(1+s)}$	1	0.2	1.0	-	-	-	-
	2	0.4	2.0	-	-	-	-
	3	0.6	3.0	-	-	-	-
	0	0	0	0	0	0	0
$\frac{20}{(1+s)^2}$	1	1.38	1.87	1.78	1.90	1.77	1.88
	2	1.78	3.06	2.33	3.07	2.31	3.09
	3	2.09	4.12	2.75	4.09	2.72	4.14
	0	3.70	3.28	3.63	3.18	3.63	3.18
$\frac{20}{(1+s)^3}$	1	4.33	4.64	4.27	4.53	4.31	4.61
	2	4.80	5.71	4.74	5.56	4.83	5.83
	3	5.19	6.65	5.13	6.44	5.26	6.84

BIBLIOGRAPHY

1. Adams, S. and M. Holmes (1977) Determining time constants for heating and cooling coils, BSRIA Technical Note TN6/77, Building Services Research and Information Association, Bracknell, UK.

2. Anon. (1977) Microprocessors in central energy controls for building. *Electrical Review*, 201, 9, pp.21-22, 10 June.

3. Anon. (1977) New fluidics operated control for variable-flow air systems. *Heating and Air-conditioning Journal*, Oct., 47 (549), 20-23.

4. Anon. (1978) Micros poised to take control in the home. *Electronics and Power*, May, p.336.

5. ASHRAE (1973) Handbook-Systems, Chapter 34, Automatic Control. *American Society of Heating, Refrigeration and Airconditioning Engineers*, New York.

6. Barney, G.C. and S.M. dos Santos (1977) *Lift Traffic Analysis Design and Control*, Peter Peregrinus, Stevenage, UK.

7. Barney, G.C. and V.L.B. de Freitas (1979) Control of building engineering services, Control Systems Centre reports 438, 443, 447, 451, 453 and 456, Control Systems Centre, UMIST, Manchester, UK.

8. Barney, G.C. and V.L.B. de Freitas (1980) A distributed data acquisition and control system for building engineering services, IEE International Conference on Effective Use of Electricity in Buildings, London, May.

9. Baxter, A.J. and A.L. Longworth (1974) The thermal response of room thermostats. *Heating and Ventilating Engineer*, pp.103-106.

10. Bondi, H. (1980) Intelligent controls in energy. *Electronics and Power*, pp. 87-93, Jan.

11. Boyar, R.E. (1962) Room temperature dynamics of radiant ceiling and air-conditioning systems. *ASHRAE Journal*, 4, No. 12, pp.59-63.

12. British Standard 799: 1964, Oil burning equipment; Part 2, Vaporizing burners and associated equipment.

13. British Standard 1042: Parts 1-3, Methods for the measurement of fluid flow in pipes.

14. British Standard 1250: 1965, Domestic appliances burning town gas, Part 4, Space heating appliances.

15. British Standard 1339: 1965, Definitions, formulae and constants relating to the humidity of the air.

16. British Standard 1523: 1967, Glossary of terms used in automatic controlling and regulating systems, Part 1: Process and kinetic control.

17. British Standard 1904: 1964, Industrial platinum resistance thermometer elements.

18. British Standard 2643: 1955, Glossary of terms relating to the performance of measuring instruments.

19. British Standard 2765: 1969, Dimensions of temperature detecting elements and corresponding pockets.

20. British Standard 2841: 1957, General purpose wet and dry bulb hygrometers.

21. British Standard 2842: 1975, Whirling hygrometers.

22. British Standard 3955: 1967, Section 2F: Room thermostats.

23. British Standard 4740: 1971, Method of evaluating control valve capacity, Part 1, Incompressible fluids.

24. British Standard 4937: Parts 1-7, International thermocouple reference tables.

25. British Standard 5169: 1975, Fusion-welded steel air receivers.

26. British Standard 5248: 1975, Aspirated hygrometers.

27. British Standard 5384: 1977, Guide to the selection and use of control systems for HVAC installations.

28. British Standard 5720: 1979, Code of Practice for mechanical ventilation and airconditioning of buildings.

All the above Standards are available from BSI, 2 Park Street, London, W1A 2BS, UK.

29. Brown, E.J. and J.R. Fellows (1957) Pressure losses and flow characteristics of multiple-leaf dampers. *Heating, Piping and Airconditioning*, 29, pp.119-125, Aug.

30. Brown, E.J. (1960) How to select multiple-leaf dampers for proper air-flow control. *Heating Piping and Airconditioning*, 32, pp.167-178, April.

31. Burberry, P.J., K.M. Letherman and D.A. Valeri, (1979) Prediction techniques for seasonal energy consumption with special reference to control systems and installation performance. Presented at the 2nd International CIB symposium on energy conservation in the built environment, Copenhagen, June.

32. Burberry, P.J. and A. Aldersey-Williams (1978) A *Guide to Domestic Heating Installation and Controls*, Architectural Press, London.

33. Caldwell, W.I., G.A. Coon and L.M. Zoss (1959) *Frequency Response for Process Control*, McGraw Hill, New York.

34. Champion, B. and F.M.H. Taylor (1960) Control of heating and ventilating systems. *Journal of the Institution of Heating and Ventilating Engineers*, 27, pp.359-63, Mar.

35. CIBS Guide (1981), Section B.11, Automatic Control, Chartered Institution of Building Services, London.

36. CIBS (1973) Commissioning code, series C, Automatic control systems, Chartered Institution of Building Services, London.

37. CIBS (1978) Symposium: Automatic controls for building services, Chartered Institution of Building Services, London.

38. Cohen, G.H. and G.A. Coon (1953) Theoretical considerations for retarded control. *Trans. of the American Society of Mechanical Engineers*, 75, pp.827-834.

39. Collister, A, (1979) Centralised control. *Journal of the Chartered Institution of Building Services*, 1, pp.41-42, May.

40. Cox, A.E. and W. Hunt (1960) A guide to the application of three-way valves. *Journal of the Institution of Heating and Ventilating Engineers*, 28, pp.305-310, Jan.

41. Coxon, W.F. (1960) *Temperature Measurement and Control*, Heywood, London.

42. Crommelin, R.D. (1978) Simulating the dynamic behaviour of a room. *Gesundheits Ingenieur*, 99 (4), pp.106-110, April (in German).

43. Croome-Gale, D.J. and B.M. Roberts (1975) *Airconditioning and Ventilation of Buildings*, Pergamon Press, Oxford.

44. Danter, E. (1974) Heat exchanges in a room and the definition of room temperature. *Building Services Engineer*, 41, pp.232-245.

45. D'azzo, J.J. and C.H. Houpis (1966) *Feedback Control System Analysis and Synthesis*, McGraw Hill, New York (2nd Ed.)

46. Delta Metal Co. (1979) Introduction to the design of shape memory effect actuators, Delta Memory Metal Co., Ipswich, UK.

47. Dept. of Energy (1977) Fuel Efficiency booklet No. 10: Controls and energy savings.

48. Di Stefano, J.J., A.R. Stubberud and I.J. Williams (1967) *Feedback and Control Systems*, Schaum outline series, McGraw Hill, New York.

49. Douce, J.L. (1963) *The Mathematics of Servomechanisms*, English Universities Press, London.

50. EEUA (Rev. 1971) A guide to the selection of automatic control valves, Engineering Equipment Users Association, London.

51. Fielden, C.J. and D.S. Laister (1978) Review of trends in the central supervising control of building services, IEE 2nd International Conference on centralised control systems, London, pp.24-28, Mar.

52. Fisher Controls Co. (1977) *Control Valve Handbook*, Fisher Controls Co., Marshalltown, Iowa, USA (2nd Ed.).

53. Fitzgerald, D. (1969) Room thermostats, choice and performance. *Journal of the Institution of Heating and Ventilating Engineers*, 37, p.127.

54. Flugge-Lotz, I. (1968) *Discontinuous and Optimal Control*, McGraw Hill, New York.

55. Galehouse, D.W. (1974) Specification and evaluation of building automation systems. *Trans. of the American Society of Heating, Refrigeration and Air-conditioning Engineers*, 80, 1, pp.393-400.

56. Gille, J.C., M.J. Pelegrin and P. Decaulne (1959) *Feedback Control Systems*, McGraw Hill, New York.

57. Griffiths, R. (1951) *Thermostats and Temperature Regulating Instruments*, C. Griffin, London.

58. Haines, J.E. (1953) *Automatic Control of Heating and Airconditioning*, McGraw Hill, New York.

59. Haines, Roger W. (1977) *Control Systems for Heating, Ventilating, and Air-conditioning*, Van Nostrand Reinhold, New York (2nd Ed.).

60. Hanby, V.I. (1976) The stability of a proportionally controlled heating system. *Building Services Engineer*, 43 (1), pp.208-211, Jan.

61. Harriott, P. (1964) *Process Control*, McGraw Hill, New York.

62. Harrison, H.L., W.S. Hansen and R.E. Zelenski (1968) Development of a room transfer function model for use in the study of short-term transient response. *Trans. of the American Society of Heating, Refrigeration and Airconditioning Engineers*, pp. 25-29, Mar.

63. Healey, M. (1967) *Principles of Automatic Control*, English Universities Press, London.

64. Honeywell (1958) Engineering Manual of Automatic Control, Minneapolis-Honeywell Regulator Co., Minneapolis, Minnesota, USA.

65. Honeywell (1977) Delta 1000 and 2000 data sheets, from number series 74-1700, Honeywell, Minneapolis, Minnesota, USA.

66. Jackson, T.B. (1971) The theory and economics of optimum start programming of heating plant. *Journal of the Institution of Heating and Ventilating Engineers*, 39, May, June.

67. Jackson, T.B. (1975) Energy conservation by improved control in industrial and commercial buildings, paper presented to the European Building Research Association, Sept.

68. Jones, W.P. (1968) Automatic controls 1-8, series of articles. *Steam and Heating Engineer*, Mar.-Oct.

69. Jones, W.P. (1973) *Air Conditioning Engineering*, Edward Arnold, London (2nd Ed.).

70. Junker, b. (1974) *Klimaregelung*, R. Oldenbourg Verlag, Munich (in German).

71. Kaye, G.W.C. and T.H. Laby (1973) *Tables of Physical and Chemical Constants*, Longmans, London (14th Ed.).

72. Kinzie, P.A. (1973) *Thermocouple Temperature Measurement*, J. Wiley, New York.

73. Kostyrko, K. (1969) Optimization of the dynamic properties of dew-point lithium chloride probes. *Journal of the Institution of Heating and Ventilating Engineers*, 37, pp.159-167.

74. Kutz, M. (1968) *Temperature Control*, J. Wiley, New York.

75. Leivers, M.F. and K.M. Letherman (1978) The extension characteristics of some materials used for humidity sensors. *Building Services Engineer*, 45, pp.205-210.

76. Lenz, H. (1964) Dynamics of controlled systems in air conditioning, Dissertation, Karlsruhe Technische Hochschule (in German).

77. Letherman, K.M. (1980) Analytical methods for thermostatic on/off controls. *Building Services Engineering Research and Technology*, 1, No. 1.

78. Letherman, K.M. (1980) Steady state responses in proportional control of room temperature, *Building Services Engineering Research and Technology*, 1, No. 4.

79. Letherman, K.M. (1976) Thermostatic control of room temperature, Ph.D. Thesis, Dept. of Building, UMIST, Manchester, UK.

80. Macklen, E.D. (1979) *Thermistors*, Electrochemical Publications, Ayr, Scotland.

81. Masoneilan (1977) Handbook for control valve sizing, Masoneilan Ltd., Middlesex, UK.

82. Miles, V.C. (1965) *Thermostatic Control*, G. Newnes, London.

83. Murray, P. (1977) Environmental control systems. *The Heating and Air Conditioning Journal*, pp.22-26, Mar.

84. Pearson, G.H. (1978) *Valve Design*, Mechanical Engineering Publications, London.

85. Pepper, J.S. and C.H. Smith (1967) Dynamic characteristics of room heaters, Laboratory Report 46, HVRA (now the Building Services Research and Information Association), Bracknell, UK.

86. Powell, D.H. (1962) Basic principles of control application. *Journal of the Institution of Heating and Ventilating Engineers*, pp.353-375, Jan.

87. Prentice, T.C. (1978) Controls in non-residential buildings. *Building Services Engineer*, 46 (3), pp.55-63, June.

88. Profos, P. and P. Hemmi (1965) The dynamics of air conditioning control. *New Techniques*, 7, A2, pp.49-56 (in German).

89. Philpott, S.R. (1977) The microprocessor and microcomputer in building services, M.Sc. Dissertation, Dept. of Building, UMIST, Manchester, UK.

90. Redmond, J. and D. Fitzgerald (1970) The performance of thermostatic radiator valves. *Journal of the Institution of Heating and Ventilating Engineers*, 38, pp.124-132.

91. Roberts, B.M. (1976) Monitoring and control of all the services in modern buildings. *Electrical Times*, 15 April.

92. Roberts, B.M. (1979) Drake and Scull Technical Report No. 5; Building Automation Systems, Drake and Scull Engineering Ltd., London.

93. Roots, W.K. (1969) *The fundamentals of Temperature Control*, Academic Press, London.

94. Royds, R. (1951) *The Measurement and Control of Temperature in Industry*, Constable, London.

95. Satchwell (1971) The design and performance of room thermostats, Technical leaflet No. 920.01, Satchwell Control Systems Ltd., Slough, UK.

96. Satchwell (1976) Digital Management System DMS2400, Automated Command Centre, Satchwell Control Systems Ltd., Slough, UK.

97. Sauter (1978) Sauter Automation EY1200, Building Automation and Power Management System, Fr. Sauter SA, Basle, Switzerland.

98. Sauter, Technical Bulletins, series, Sauter Automation Ltd., Slough, UK.

99. Schwarzenbach, J. and K.F. Gill (1978) *Systems Modelling and Control*, Arnold, London.

100. Shih, J.Y. (1975) Energy conservation and building automation, *Trans. of the American Society of Heating, Refrigeration and Airconditioning Engineers*, 81, 1, pp.419-435.

101. Shinners, S.M. (1964) *Control System Design*, J. Wiley, New York.

102. Spencer-Gregory, H. and E. Rourke (1957) *Hygrometry*, Crosby-Lockwood, London.

103. Southerton, R. (1971) An analysis of the control behaviour of some heating and airconditioning systems, M.Sc. Dissertation, UMIST, Manchester, UK.

104. Stanley, E.E. (1964) A combined theoretical and analogue computer study of on/off thermostats, Laboratory Report No. 20, HVRA (now the Building Services Research and Information Association), Bracknell, UK.

105. Telcon (1978) Designers handbook for the selection and use of thermostatic bimetals, Telcon Metals Ltd., Crawley, Sussex, UK.

106. Thal-Larsen, H. (1960) Dynamics of heat exchangers and their models. *Trans. ASME Journal of Basic Engineering*, 82, pp.489-504, June.

107. Trane (1970) *Introduction to Control Application*, The Trane Co., Wisconsin, USA.

108. Tsypkin, J.S. (1955) *Teorija Relejnykh Sistem Automaticheskovo Regulirovanija*, Gostekhizdat, Moscow (in Russian), see ref. 56, Gille, above.

109. Trethowen, H.A. (1970) Control valve performance. *Journal of the Institution of Heating and Ventilating Engineers*, 37, pp.232-240.

110. Whalley, R. (1977) Microprocessors in control systems. *Electrical Review*, 200, 22, p.44, June.

111. Wolsey, W.H. (1967) *Electrical Controls for Heating and Air Conditioning*, VDI Verlag, Dusseldorf, 2 vols. (in German).

112. Wolsey, W.H. (1967) The operation of steam control valves for heating installation. *Journal of the Institution of Heating and Ventilating Engineers*, 35, pp.35-50, May.

113. Wolsey, W.H. (1971) A theory of three-way valves. *Journal of the Institution of Heating and Ventilating Engineers*, 39, pp.35-51, May.

114. Wolsey, W.H. (1975) *Basic principles of Automatic Control*, Hutchinson Educational, London.

115. Zermuehlen, R.O. and H.L. Harrison (1965) Room temperature response to a sudden heat disturbance input. *American Society of Heating, Refrigeration and Airconditioning Engineers Journal*, pp.25-29, Mar.

116. Ziegler, J.G. and N.B. Nichols (1942) Optimum settings for automatic controllers. *Trans.* ASME, 64, pp.759-763.

PROBLEMS

1. Show that the following transforms:

   (a) $(2s+a)/s(s+a)$,

   (b) $(s+3)/(s+4)(s-2)$,

   (c) $1/(s^2+4)(s^2+9)$,

   have the following partial fraction expansions:

   (a) $1/s + 1/(s+a)$,

   (b) $1/6(s+4) + 5/6(s-2)$,

   (c) $1/5(s^2+4) - 1/5(s^2+9)$,

   and the following inverse transformations:

   (a) $1 + e^{-at}$,

   (b) $(e^{-4t} + 5e^{2t})/6$,

   (c) $(1/10)\sin 2t - (1/15)\sin 3t$.

2. Show that the final values of the inversions of

$$10/((s+1)(s^2+2s+3)) \text{ and } K/s((s+1)(s+2)(s+5)+K)$$

   are 0 and $K/(10+K)$ respectively.

3. Show that

$$L^{-1}(10/(s+1)^2(s^2+2s+2)) = 10e^{-t}\,(t - \sin t).$$

4. If $F(s) = 13s(s^2+4s+13)$, show that the initial and final values of $f(t)$ are 0 and 1 respectively.

5. Show that the Laplace transform of the function

$$f(t) = 0 \quad (t < 2)$$
$$= 1 \quad (2 < t < 3)$$
$$= 0 \quad (3 < t)$$

is $(e^{-2s} - e^{-3s})/s$.

6. The curve in Fig. P.1 shows the step response of a simple first-order thermometer bulb of time constant T seconds. The straight line AB is tangent to the curve at C. Show that $t_1 = T$.

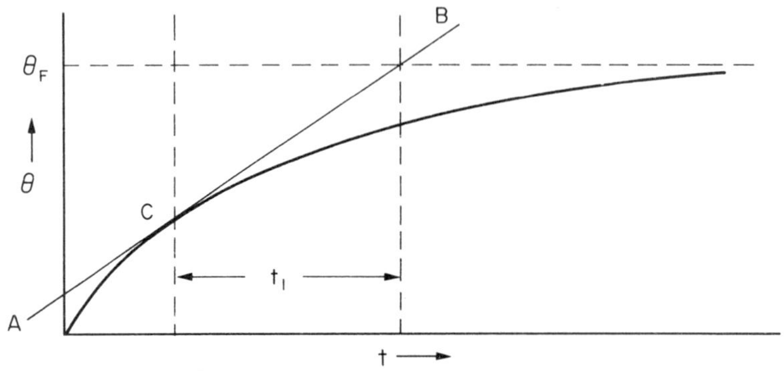

Fig. P.1.

7. A mercury-in-glass thermometer has a cylindrical bulb 1 in. long and $\frac{1}{4}$ in. diameter. The heat-transfer coefficient between the bulb and still water may be taken as 30 Btu/hft^2 $^{\circ}$F. Find the thermal time constant of the bulb in still water, mentioning all assumptions made.

(16.5 sec)

8. An industrial resistance thermometer was suddenly inserted in a large vessel of liquid after having been at a steady temperature of 20°C. The subsequent readings are tabulated below. Use them to determine the time constant of the thermometer in as many different ways as possible.

t, sec	0	4	8	12	16	20	30	40	50	60	70	80	120
θ, $^{\circ}$C	20	57	89	108	131	145	176	191	204	210	215	217	220

9. A thermometer bulb of 1 cm outside diameter has a thermal time constant of 5.0 seconds in water at 60°C flowing at 1 m/s.

(a) If the water temperature undergoes a step change from 50°C to 70°C, find:

(i) the time taken for the reading to reach 60°C,

(ii) the reading after 15 seconds,

(iii) the time taken to reach 62.64°C.

(b) Estimate the time constant at 60°C:

(i) in still water,

(ii) in still air,

(iii) in water flowing at 4 m/s,

(iv) in air flowing at 10 m/s.

DATA:

For free convection: $Nu = 0.53\ (Gr \cdot Pr)^{0.25}$

For forced convection: $Nu = 0.26\ (Re)^{0.6}\ (Pr)^{0.3}$.

<u>Water at 60°C:</u> $Pr = 3.0$, kinematic viscosity $= 4.78 \times 10^{-7}\ m^2/s$.

$k = 0.65\ W/m^oC$, $Gr = 2.24 \times 10^{10} \cdot d^3 \cdot \Delta\theta$.

<u>Air at 60°C:</u> $Pr = 0.70$, kinematic viscosity $= 1.88 \times 10^{-5}\ m^2/s$.

$k = 2.88 \times 10^{-2}\ W/m^oC$, $Gr = 0.82 \times 10^{8} \cdot d^3 \cdot \Delta\theta$.

Assume $\Delta\theta = 10^oC$, all heat transfer by convection.

(a: 3.47 s, 69°C, 5.0 s)
(b: 47 s, 110 min, 2.2 s, 400 s)

10. Figure P.2 shows a simple model of an electric storage heater. The rate of heat input q is 750 W. The rate of heat loss per °C temperature difference K is 10 W/°C. The thermal capacity C_1 is 8×10^4 J/°C. θ_2 is ambient temperature.

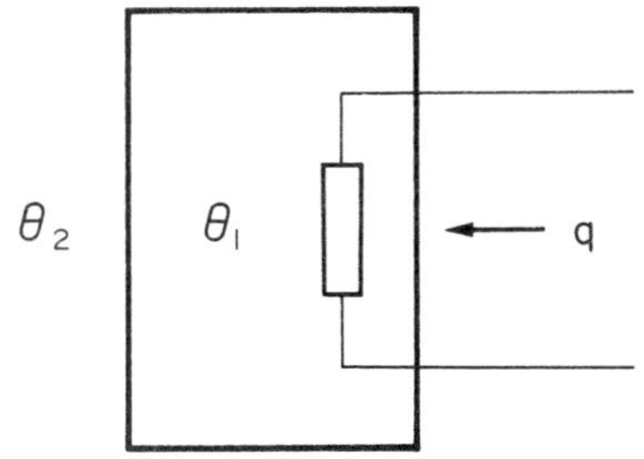

Fig. P.2.

(a) Find the differential equation relating q, θ_1 and θ_2.

(b) Show that the transfer function relation q and θ_1 is $1/(K + C_1 s)$.

(c) Find the equation describing the time response of θ_1 to a sudden switching on of q.

(d) Find the initial rate of temperature rise θ_1 and confirm that this is equal to q/C_1.

(e) Find the temperature rise 1 hour after switch-on. (27.2°C)

(f) Find the temperature rise after a long time, and confirm that this is equal to q/K.

11. The temperature of a well-stirred water tank is to be controlled by an electric proportional controller of gain K. The steady-state heat loss of the tank is 50 W/°C, and it contains 10 kg of water.

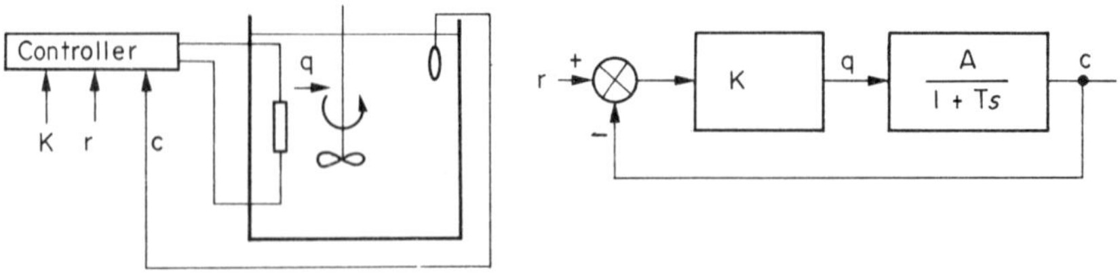

Fig. P.3.

Initially the reference temperature r = tank temperature, c = 15°C.

At time t=0 the reference input r is suddenly changed from 15°C to 25°C and left there. After a long time the offset is found to be 1°C.

(a) What are the values of T, A and K?

(b) Is c(t) oscillatory for any value of K?

(c) Find the offset and the final value of c(t) if K is (i) doubled, (ii) halved.

(d) Sketch c(t) for K = 1/A, 4/A, 9/A, 49/A.

(e) If K = 9/A, find c(168).

(840 s, 0.02°C/W, 450 W/°C, 0.53°C, 24.47°C, 1.82°C, 23.18°C, 22.78°C)

12. For the system shown in Fig. P.4 with a certain value K_1 of the gain K, the

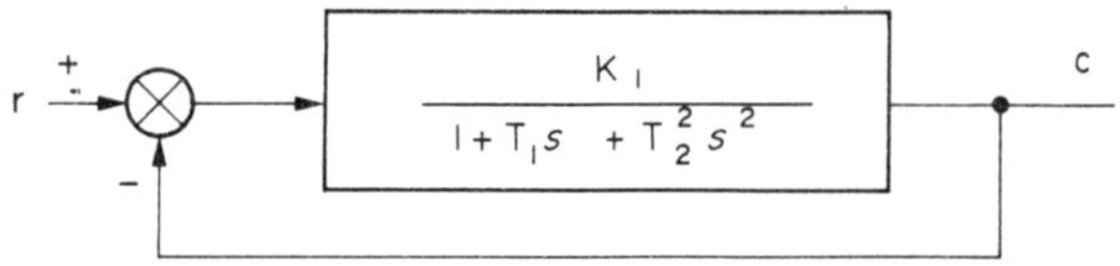

Fig. P.4.

input r was suddenly changed from 120°C to 170°C and left there. The output c was observed to change from 120°C and exhibit a damped oscillation, eventually settling at 160°C. The highest temperature reached was 172.5°C, at 80 seconds after the step change in r, as shown in Fig. P.5.

(a) Find K_1, T_1 and T_2.

(b) Find the value to which K should be set in order that c(t) as $t\to\infty$ should approach 170°C as closely as possible, but that c(t) should never exceed 180°C.

(a: $K_1 = 4$, $T_1 = 82.5$ s, $T_2 = 53.5$ s)

(b: $K = 5.75$, $\zeta = 0.297$)

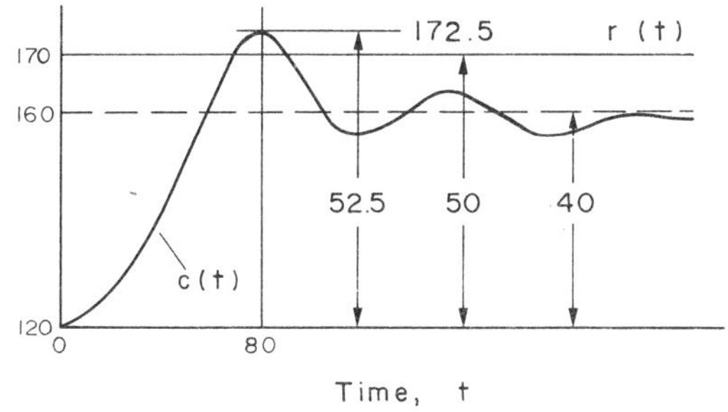

Fig. P.5.

13. In the controlled system shown in Fig. P.6, the time constants T_1 and T_2 are 1100 seconds and 100 seconds. The constant A is 0.05°C/W.

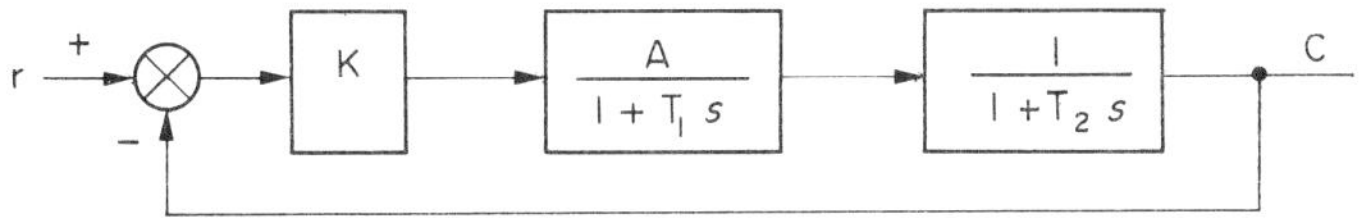

Fig. P.6.

(a) Find the value of K at which C(t) is just oscillatory for a step input r(t).

(b) Find the value of K to give 20% overshoot to a step input, and the resultant values of W and ζ.

(c) Find the percentage offset in cases (a) and (b).

(d) What are the order, type and rank of this system?

(e) Sketch an s-plane plot of the roots of the characteristic equation for various values of K.

(45.6 W/°C; 296 W/°C, 8.6 min, 0.46; 30.5%, 6.3%; 2, 0, 2)

14.

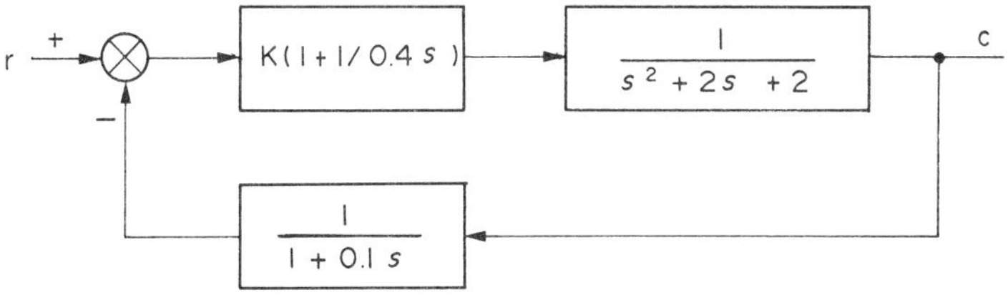

Fig. P.7.

(a) What are the order, type and rank of the system shown in Fig. P.7?

(b) Find the value of K at which the system becomes unstable.

15. A room air-temperature sensor consists of a coil of "BALCO" wire wound on an open plastic former. The length of wire used is 1 metre and it is 36 s.w.g.

The room temperature, which has been steady at 15°C for a long time, suddenly starts to rise at a constant rate of 0.3°C per second.

(a) Obtain the transfer function of the sensor, relating air and wire temperatures, and hence the transfer function relating air temperature and wire resistance.

(b) Determine the equation describing the time variation of the sensor resistance.

(c) When the air temperature reaches 20°C, what is the wire temperature? What is the time delay between 20°C being reached by the room air and 20°C being reached by the wire?

(18.5°C, 7.35 sec)

(d) Upon reaching 25°C, the air temperature stops rising and remains constant. How long after the start of the original temperature rise will the detector reach the resistance corresponding to 24.9°C?

(57.6 sec)

(e) Sketch the temperature/time curves for air and wire.

DATA: 36 s.w.g. = 1.93×10^{-4} m diameter.

Specific heat of wire = 400 J/kg°C.

Density of wire = 8000 kg/m^3.

Resistivity of wire = 7×10^{-8} ohm metre at 20°C.

Temperature coefficient of resistance of wire = 4.3×10^{-3} °C^{-1} related to 0°C.

Heat-transfer coefficient, wire to air = 20 W/m^2 °C.

Assume all the wire surface to be exposed to air.

16. In the system shown in Fig. P.8, the gains K_1 and K_2 control the amounts of

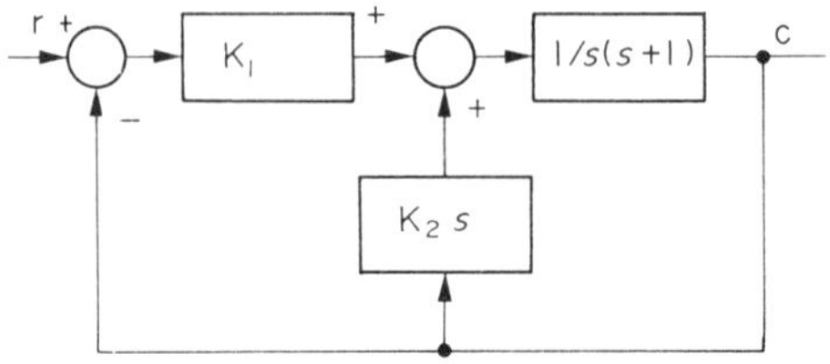

Fig. P.8.

error and error rate fed to the process.

(i) Find K_1, K_2 to give a closed-loop transfer function with a damping ratio of 0.6 and a damped natural frequency of 10 rad/s.

(156, 14)

(ii) What is then the overshoot for $r(s) = 1/s$?

(9.5%)

(iii) What are the times of first and second crossovers of the response?

(0.22 s, 0.53 s)

17. A two-tank level control system behaves substantially with a second-order type of response. In order to avoid spillage, overshoot of the response to a unit step input must not exceed 10%. In order to fulfil process requirements, the rise time of this response must not exceed 12.5 seconds. Find the necessary values of ζ and ω_n, and the resulting values of 5% settling time and peak time.

18.

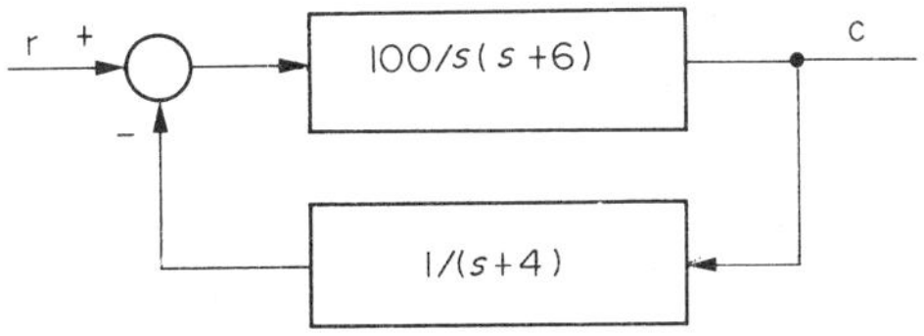

Fig. P.9.

The system shown in Fig. P.9 has a third-order CLTF. Find the time function of C in response to a unit step at r. Plot c(t) for $0 < t < 3$. Show that c(t) is closely approximated by a second-order response with $\zeta = 0.21$, $\omega_n = 3.4$. Compare the overshoot values, rise times and settling times of the two responses.

19. For each of the characteristic equations given below determine the conditions on K for stability.

(i) $as^3 + bs^2 + cs + K = 0$ $(K < b\,c/a)$,

(ii) $as^3 + bs^2 + Ks + c = 0$ $(K > a\,c/b)$,

(iii) $as^4 + bs^3 + cs^2 + ds + K = 0$ $(K<((cd/b) - (ad^2/b^2)))$.

20. Determine the stability of each of the following cases:

(i) $\text{OLTF} = s^4 + 3s^3 + s^2 + 6s + 1$ (unstable).

(ii) Characteristic function $= s^3 + 3s^2 + 4s + 12$ (stable).

(iii) Characteristic equation: $s^5 + s^4 + 4s^3 + 4s^2 + 2s + 1 = 0$ (unstable).

(iv) Characteristic equation: $s^4 + 2s^3 + 11s^2 + 18s + 18 = 0$ (stable).

21. Determine the conditions on K for stability for each of the following characteristic equations:

(i) $s^4 + 7s^3 + 15s^2 + (25 + K)s + 2K = 0$ (K < 28.1),

(ii) $(s + 2)(s^2 + 4s + 8) + K = 0$ (K < 80),

(iii) $s^4 + 6s^3 + 11s^2 + 6s + (1 + K) = 0$ (K < 9),

(iv) $s^5 + 2s^4 + 10s^3 + (8 + K)s^2 + 9s + 4K = 0$ (stable for 0 < K < 2).

22. For the general characteristic equation

$$a_0 s^n + a_1 s^{n-1} + a_2 s^{n-2} + \dots + a_{n-2} s^2 + a_{n-1} s + a_n = 0.$$

Show that dividing through by a_n and making the substitution $s = V_p$ (where V is a positive constant) does not alter the information on stability given by the Routh Hurwitz criterion.

23. Determine the number of roots with positive real parts for each of the following equations:

(i) $100s^5 + 1100s^4 + 9 \times 10^3 s^3 + 1.05 \times 10^5 s^2 + 1.2 \times 10^6 s + 10^7 = 0,$

(ii) $s^5 + 1.1s^4 + 0.9s^3 + 10^{-2}s^2 + 10^{-4}s + 10^{-6} = 0.$

(Two, none)

24. Find the value of A for which the equation

$$X^3 + 5X^2 + 4X + A = 0$$

has (i) a pair of purely imaginary roots, (ii) one real root $\geqslant$ 0. Find the values of the roots in each case.

25. Sketch the Root Locus of the system which has the OLTF: $G(s)\,H(s) = K/s(s+4)(s+5)$ and determine the values of K at which the system becomes (a) oscillatory and (b) unstable. (13.1, 180)

26. Sketch the Root Locus plot for a system with OLTF: $K/s(s^2+4s+8)$. Find the limiting value of K for stability, and find the value of K which gives an equivalent second-order damping factor of 0.4

27. (a) Show that a system with OLTF:$K/((s+1)(s-1)(s+4)^2)$ is unstable for all values of K.

(b) In order to stabilise the system, a compensator is added, which changes the OLTF to:

$$K(s+A)/((s+1)(s-1)(s+4)^2).$$

Draw the root locus plots for A = 0.5, 1, 2. Find the ranges of K in each case to give stable response.

28.

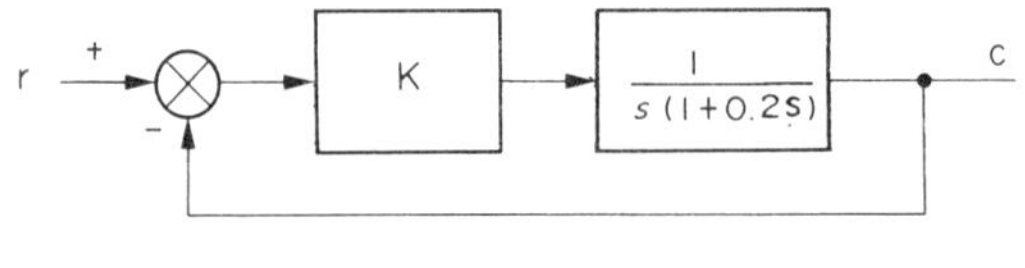

Fig. P.10.

(a) Plot the Root Locus of the controlled system shown in Fig. P.10, for various values of K. For K = 100 determine the steady-state error when r(t) is a unit ramp function, and estimate the transient behaviour.

(b) If the controller now has derivative action added to it, giving a controller transfer function $K(1+0.1s)$, draw the Root Locus plot and estimate the effect on the steady-state error and transient response for K = 100.

29.

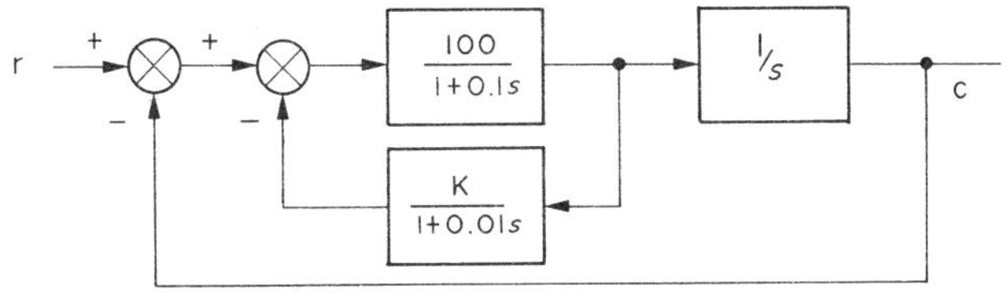

Fig. P.11.

The block diagram of a position control servomechanism is shown in Fig. P.11. Draw the Root Locus for this system with K as the variable parameter. Determine the value of K corresponding to a damped natural frequency of 20 Hz.

(0.178)

30. Plot the asymptotic log magnitude and phase diagrams for each of the following open-loop transfer functions, and determine whether or not they are stable:

$$\frac{2(1+0.3s)}{s(1+0.1s)(1+0.4s)}, \quad \frac{5}{s(s^2+16s+400)}.$$

31. Plot the Nyquist diagram for the system whose OLTF is

$$K/s(1+s)(1+2s).$$

Find the value of K to give 6 dB gain margin, and find the corresponding phase margin.

32. Show that the system whose OLTF is

$$1000/(1+s)^2(1+1000s)$$

is stable and find the phase and gain margins.

33. Construct the asymptotic and the exact Bode plots for the following transfer functions:

$$\frac{10(1+s)}{s^2(1+s/4+s^2/16)}, \quad \frac{(4s+1)}{s(1+2s)(1+8s)}.$$

34. For the system with OLTF: $GH = \dfrac{K\exp(-s)}{(1+s)^2}$

(a) Find the limiting value of K for stability using the first-order Padé approximation for the pure delay term.

(b) Repeat (a) using the second-order approximation.

(c) Compare the limits given by (a) and (b) with the result given by an exact solution.

35. Draw the frequency response of the pure delay $\exp(-2s)$ using both polar plots and logarithmic plots. Compare these plots with those for the first- and second-order Padé approximations.

36. Sketch the general shape of polar plot for each of the transfer functions given below:

$$\frac{K}{s(1+Ts)}\,, \quad \frac{Ke^{-T_1 s}}{1+T_2 s}\,, \quad \frac{K}{(1+Ts)^4}\,, \quad \frac{K(1+Ts)}{(1+2Ts)}\,.$$

37. A temperature-control system consists of a sensor of time-constant T_1, a proportional controller of gain K and a heater element of time-constant T_2. In an open-loop frequency response test, the response sine wave was found to lag 90° behind the input wave at a frequency of 0.0159 Hz. At this frequency the response amplitude is 4 times that of the input. At very low frequencies the response amplitude is 10 times that of the input. Find T_1, T_2 and K.

(5s, 20s, 10)

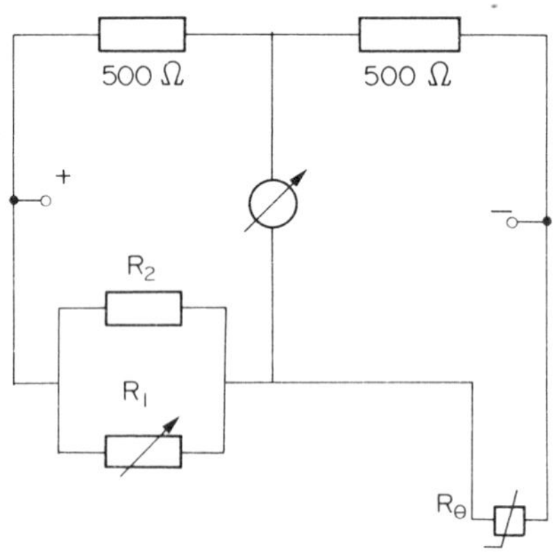

Fig. P.12.

38. Figure P.12 shows a resistance thermometer circuit for measuring the temperature rise in a Bomb Calorimeter. The bridge is used in the null-balance mode. The two 500-Ω resistors are fixed, and R_1 is variable manually between 10 kΩ and 15 kΩ. The detector bulb has a resistance of 450 Ω at 0^oC and a temperature coefficient of resistance of $6.1 \times 10^{-3}\ {}^oC^{-1}$ related to 0^oC.

The resistor R_2 must be chosen so that when R_1 is 10 kΩ the bridge is in balance at $\theta = 20^oC$. The bridge is balanced manually.

(a) Calculate R_2.

(b) Calculate the range of temperature that can be measured by this arrangement.

39. The resistance of a certain thermistor at various temperatures is given in the table below. The same thermistor has its characteristic linearised by having a fixed resistor of 25 kΩ placed in parallel with it.

Draw resistance/temperature curves for the linearised and non-linearised cases on the same graph. For each, find the temperature coefficient at 20°C related to 0°C.

Temperature,	°C	10	15	20	25	30
Resistance,	kΩ	56	36	25	17	11

40. A type K thermocouple of 2 Ω resistance is connected by copper leads of negligible resistance to a galvanometer and a Manganin swamp resistor. The galvanometer has a copper coil of resistance 60 Ω at 20°C, and deflects fully with a voltage of 6 mV across it. The error in reading due to ±5°C ambient temperature variation at the galvanometer must not exceed $\pm \frac{1}{2}$%.

(a) Find the necessary value of the swamp resistor.

(b) Find the range of temperature for which the instrument is suitable, to the nearest 100°C.

Temperature coefficient of resistance of copper, related to 0°C = 4×10^{-3} °C^{-1}.
Thermocouple e.m.f. = 4.1 mV per 100°C.

41. A liquid expansion dial thermometer is filled with mercury. The capillary and bulb are of stainless steel. The internal dimensions of the bulb are 12.7 mm i.d. by 125 mm long, and the capillary tube is 0.5 mm bore x 25 m long.

(a) Find the volume expansion when the bulb temperature rises from 100°C to 200°C.

(b) If the mean capillary temperature changes by 20°C, find the resultant error in the reading.

(c) If the size of the bulb were increased to 18 mm i.d. x 200 mm long, what would be the new error in reading in part (b)?

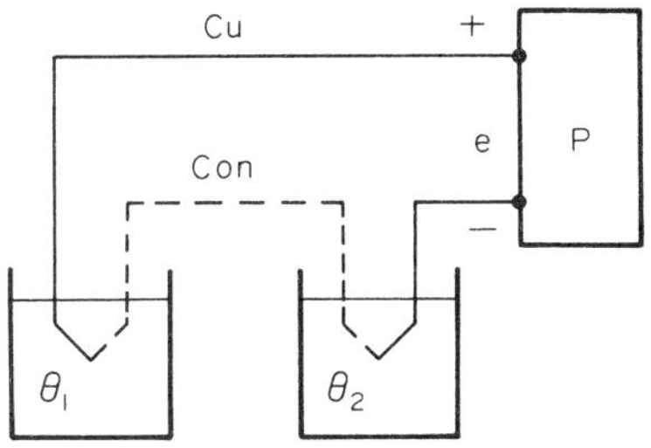

Fig. P.13.

42. In Fig. P.13, P represents a potentiometer measuring the thermal e.m.f. e produced by the type T thermocouple. For each of the following cases, calculate the temperature θ_1, also the uncertainty limits on θ_1. See BS 4937,

part 5.

(a) θ_2 is maintained at 0^oC and e = 4834 ± 0.5 μV,

(b) θ_2 is maintained at 45.5 ± 0.1^oC and e = -2.83 ± 0.01 mV,

(c) θ_2 is 100^oC and e = + 5782 ± 1 μV.

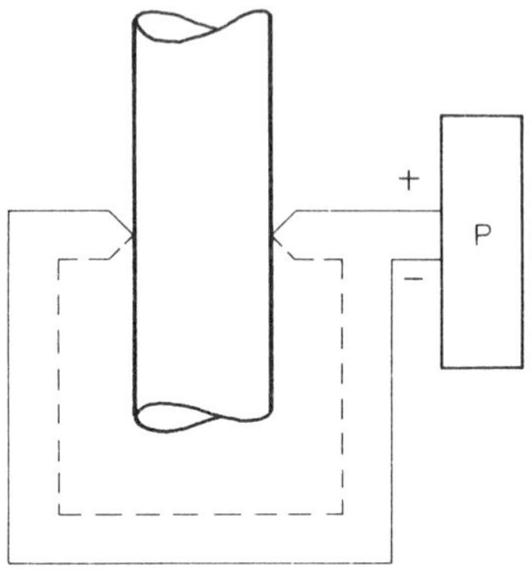

Fig. P.14.

43. Figure P.14 shows a thermocouple arranged differentially to measure the temperature difference $\Delta\theta$ between the faces of a sample under test.

An appropriate equation for the hot junction temperature θ of a thermocouple when the cold junction is at 0^oC is:

$$\theta = 2.6065 \times 10^{-2}\ E - 7.2672 \times 10^{-7}\ E^2 + 3.8576 \times 10^{-11}\ E^3 - 1.1046 \times 10^{-15} E^4$$

where E is the e.m.f. in μV.

In this case the mean temperature of the sample is known to be about 25^oC. The e.m.f. E is measured to be 124.6 ± 0.2 μV. Find $\Delta\theta$ as accurately as possible.

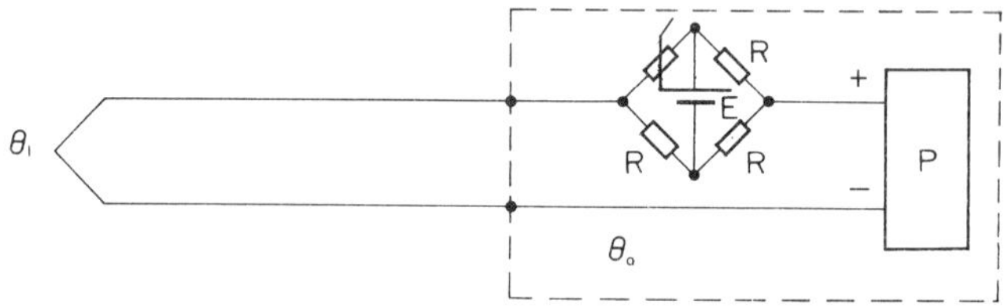

Fig. P.15.

44. Figure P.15 shows an arrangement for automatic compensation of changes in ambient temperature θ_a. Three of the resistors in the bridge are of Manganin, each of resistance R, the fourth is of Platinum, whose resistance may be taken as $100\ (1+3.92\times10^{-3}\theta_a)$ ohms. The bridge is supplied with a stable direct voltage of E volts.

Design the bridge, choosing the values of R and E so that when θ_a is 20°C the e.m.f. seen by P is that which would be seen if the cold junction were at 0°C, and at other values of θ_a, the bridge output changes automatically to compensate for the change in θ_a. Estimate the permissible variation in θ_a for acceptable compensation.

Assume the thermocouple is Chromel/Alumel, type K. See BS 4937, part 4.

45. In each of the bridges shown in Fig. P.16 the resistors marked R are fixed at 500 Ω. The resistors r are small trimming resistors of 20 Ω each. The sensing resistor is BALCO and has a resistance of 500 Ω at 20°C. Each core of the

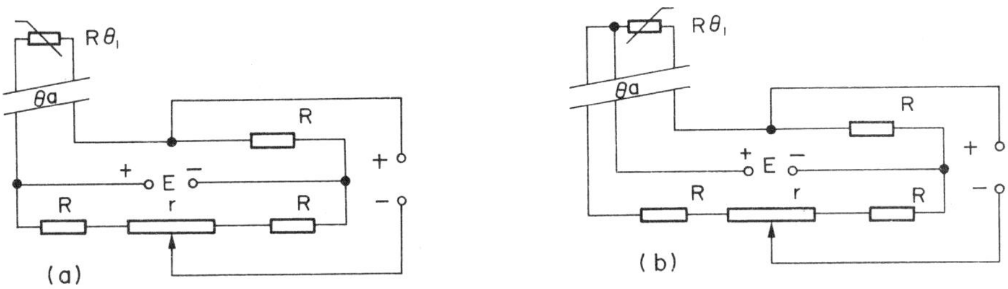

Fig. P.16.

cable in the region θ_a has a resistance of 4 Ω.

Initially $\theta_1 = \theta_a = 20°C$. The sliders are adjusted so that e is zero. For each bridge, find the output e when (a) θ_1 changes to 25°C, (b) θ_a changes to 25°C.

Assume E = 6 V.

46. The calibration of a P+D controller is to be checked experimentally. The input is subjected to a steady linear change of 10% per minute. The output undergoes an initial step change of 5% followed by a steady linear change of 2% per minute. Find the proportional gain constant K_P, the proportional band PB and the derivative action time T_D.

(0.2, 500%, 2.5 min)

47. A pneumatic P+I controller has the following settings: PB = 50%, T_i = 5 minutes. The input is subjected to a steady linear change of 0.5 psi per minute, with both the input and output starting at 3 psi and increasing.

(a) How long after the input starts to change will the output due to proportional action be equal to the output due to integral action?

(b) What is the initial rate of change of output?

(c) What will be the value of the input when the output saturates?

(10 min, 1 psi/min, 6.5 psi)

48. The temperature of a gas-heated oven is controlled by a proportional controller actuating the gas valve motor over the standard range of 3-15 psig. The sensor is a thermocouple situated in the oven.

With 100% PB the full output range of the controller corresponds to the range of temperature covered by the indicator, which is 900→1100°C.

When on manual control, the pressure at the valve motor was subjected to a step change of 1 psi. After a delay of 2 minutes the oven temperature was seen to fall at a rate of 8.55°C/min. The final change in temperature was 60°C.

Assuming linearity throughout the control loop, determine the limiting PB for stability. What PB would you set in practice?

(58%, 116%)

49. A process consists of a simple first-order lag of time-constant 4 minutes. The loop is closed by a P+I controller, and a sensor which has a distance-velocity lag of 0.5 minute. All other lags and time-constants are negligible. If the proportional gain K_p is 5, find the smallest integral action time T_i which can be set without instability.

(0.54 min)

50.

Time (Min)	0	0.5	1.0	1.5	2.0	2.5	3.0	3.5	4.0	4.5	5.0	5.5	6.0	6.5	7.0	8.0
θ (°C)	15.0	15.0	15.0	15.1	15.3	15.8	16.4	16.9	17.5	18.1	18.7	19.2	19.5	19.7	19.9	20.0

The table above gives values of the air temperature θ when the control loop is opened and a 1 psi step change applied to the control valve of a room heater. The controlled range is 10°C to 30°C.

Determine the approximate suitable values for K_p, PB, T_i and T_d if the controller is:

(a) Proportional

(b) Proportional + Integral

(c) Proportional + Integral + Derivative.

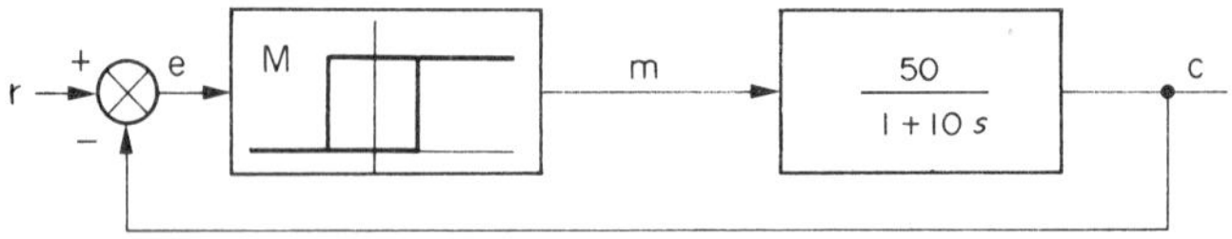

Fig. P.17.

51. A first-order process of time-constant 10 minutes is controlled by an on/off thermostat with a total hysteresis of 5°C. M = 2, r = 75°C.

Use the step-response method to find the amplitude and frequency of the controlled oscillations. Why is the Describing Function method not applicable in this case?

52. A process with transfer function $K/s(1+s)^2$ is to be controlled by one of the following relay types:

(a) On/off with zero hysteresis.

(b) On/off with hysteresis h.

For each relay type, show that the first term in the Fourier series for m(t) is given by:

(a) $\frac{2M}{\pi}\sin \omega t$, (b) $\frac{2M}{\pi}\sin(\omega t-\phi)$

where $\sin\phi = h/2X$, M = peak-to-peak output of relay

where $m(t) = \sum_{n=1}^{\infty} A_n \sin n\omega t + \sum_{n=1}^{\infty} B_n \cos n\omega t$

and $A_n = \frac{\omega}{\pi}\int_0^{2\pi/\omega} m\cdot\sin n\omega t\cdot dt$, $B_n = \frac{\omega}{\pi}\int_0^{2\pi/\omega} m\cdot\cos n\omega t\cdot dt$.

If $K = 30^{\circ}C/kW$ and M = 0.75 kW, use the Describing Function method to find the amplitude and period of the controlled oscillations in each case.

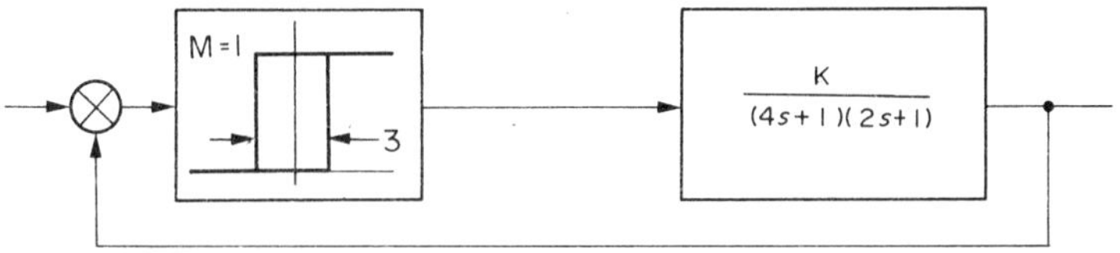

Fig. P.18.

53. Use the Describing Function method to find the minimum value of K which will just give steady oscillations for the system shown in Fig. P.18.

54. In Problem 53, if K = 20, use Tsypkin's method to find the frequency of controlled oscillations. Compare your result with that given by the Describing Function method.

55. A system $L(s) = 10/1+0.4s)(1+2s)$ is controlled by a relay with h = 0.2. Find the period of oscillations. Also determine the new value of h necessary to give a period of oscillation of 1.4. Estimate the amplitude in each case. M = 1.

56. An air-conditioned computer room has a 5-kW heater battery in the air-conditioning plant. The uncontrolled system response was tested by switching on the heater battery at full output and leaving it on while recording the response of the extract air temperature sensor.

It was observed that after a delay of 1 minute the temperature began to rise at an instantaneous rate of $1.5^{\circ}C$ per minute. The final temperature rise attained was $30^{\circ}C$.

A proportional plus integral controller is to be fitted to control the extract air temperature via the heater battery. The extract air temperature sensor is of negligible time constant.

Construct the closed-loop block diagram for the controlled system, inserting and briefly justifying the transfer function in each block. If the integral action time is set equal to half the system time constant, calculate the controller gain which gives 6 dB gain margin.

Briefly describe any rule-of-thumb method for choosing settings of two term or three term controllers. (UMIST)

57. Show that the fundamental requirement for the stability of a negative feedback system is that no root of the characteristic equation has a positive real part.

Explain how this requirement is indirectly investigated by *each* of two different stability criteria. Comment on the suitability of each criterion for systems containing distance-velocity lags. (UMIST)

58. Describe the characteristics of electrical resistance temperature sensors.

Show how these sensors may be used in temperature control, including methods of lead resistance compensation, authority provision, automatic reset, sequencing control and night set-back.

An air-temperature sensor, wound from BALCO wire, has a resistance of 1000 Ω at 20°C. Design a Wheatstone bridge for proportional control, using this sensor, and with the following arrangements:

(a) Proportional band variable from 2% to 200%.

(b) Set point variable from 0 to 50°C.

(c) Night set-back by 10°C (switched).

Temperature coefficient of resistance of BALCO wire referred to 0°C is 4.3×10^{-3} per °C. (UMIST)

59. Given that the Fourier series for a square wave of amplitude M and frequency ω is

$$\frac{2M}{\pi} \sum_{n=1}^{\infty} \frac{1}{n} \sin n\omega t$$

develop the describing function for a relay with total hysteresis h.

Use this function to estimate the frequency and amplitude of oscillations when such a relay, with unity output and 3°C total hysteresis, is used to control the temperature of a system. The forward transfer function of the system is $10/(4s+1)(2s+1)$ and it has unity feedback. Comment on the accuracy of your results. (UMIST)

60. (a) State the laws of thermoelectricity and describe how they are applied to practical temperature measurement with thermocouples.

(b) Describe the purpose of the swamp resistor in a galvanometer/thermocouple circuit.

(c) Describe a method of automatic compensation for changes in ambient temperature when using a potentiometer for thermocouple measurements. Develop the equation for the compensating resistor.

(d) Describe how thermocouples may be used (i) to measure the average of the temperatures at a number of points and (ii) to increase the e.m.f. output for differential measurements. What precautions must be taken in each case? (UMIST)

61. The *describing function* N(X) for an ideal relay with zero hysteresis and output M is a real number $2M/\pi X$, where X is the amplitude of the sinusoidal input to the relay. Show how this result can be used to obtain the describing function for a relay with a total hysteresis h.

The air temperature in a room may be controlled by either of two thermostats. One thermostat has zero hysteresis and the other has 2°C total hysteresis. In each case the heater output is 3 kW and the overall transfer function of the linear part of the control loop may be taken as

$$\frac{10}{(1 + 100s)^3} \quad {}^{\circ}\mathrm{C/kW}.$$

Compare the period and amplitude of oscillations of room air temperature for each of the two cases. (UMIST)

62 (a) Describe in detail the operation of an electrical control system using a balancing relay. Show how the circuit is arranged to allow change of the set point and the proportional band and to provide upper and lower temperature limiting.

(b) Describe in detail the operation of an electronic controller which uses a Wheatstone bridge. Include the arrangements for change of the set point and the proportional band and for compensation of changes in ambient temperature. Describe the bridge arrangement for sequencing control of heating and cooling valves, with relay control of the refrigeration plant. (UMIST)

63. Figure P.19 shows a second-order process controlled by a proportional controller of gain K.

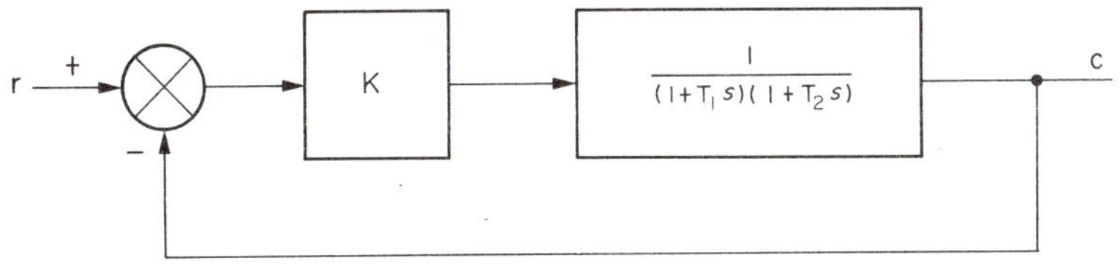

Fig. P.19.

By reference to a standard for of second-order transfer function, describe how the nature of the step response depends on the damping factor ζ and hence on K. Sketch the corresponding roots of the characteristic equation for various values of K.

If the figure represents a temperature control system, with $T_1 = T_2 = 200$ seconds, find the value of K which will give 10% overshoot in response to a step input at r. (UMIST)

64. Given that the periodicity conditions for a self-oscillating system are

$$\theta\left(\frac{\pi}{\omega}\right) = -\frac{h}{2} \quad \text{and} \quad \frac{d\theta}{dt}\left(\frac{\pi}{\omega}\right) < 0$$

develop the equations for Tsypkin's method of non-linear system analysis.

Hence show how the Tsypkin locus may be developed from the Nyquist locus of a linear transfer function and how the conditions for a periodic solution can be applied to find the frequency of oscillation. (UMIST)

65. For the general transfer function:

$$\frac{K(1 + T_a s)\ldots(1 + T_b s)\exp(-Ts)}{s^n(1 + T_1 s)\ldots(1 + T_2 s)(1 + 2\zeta s/\omega_n + s^2/\omega_n^2)}$$

develop the separate plots for approximate Bode diagrams, using straight line asymptotic approximations where possible.

Use this method to analyse the stability of the transfer function

$$\frac{10(s + 2)\exp(-5s)}{s(s + 3)} .$$

(UMIST)

66. Show that the steady sinusoidal response of a linear system represented by the Laplace transform $F(s)$ can be obtained by evaluating the function $F(j\omega)$, where ω is the frequency of interest.

Construct the Bode plots and determine the gain and phase margins for the system with the open-loop transfer function:

$$\frac{4}{(1 + s)(1 + 0.3s)^2} .$$

(UMIST)

67. Describe in detail the working characteristics of the following types of humidity sensors:

(a) Hair elements.

(b) Nylon elements.

(c) "Dewcel" lithium chloride sensors.

Comment on the relative suitability of each for humidity control in buildings. (UMIST)

68. Draw the root locus plot for the system whose open-loop transfer function is

$$GH = \frac{K}{s(s + 1)(s^2 + 7s + 12)}$$

for positive values of K.

(a) Hence determine the minimum value of K which will produce an oscillatory step response.

(b) Determine the maximum value of K which will produce a stable response.

(c) If the gain K is fixed at the mean of these two values, determine the equivalent second-order damping ratio ζ and the damped natural frequency ω_d. Justify the use of second-order parameters in this case. (UMIST)

69. (a) Sketch a Wheatstone bridge circuit suitable for proportional control with an amplifier, modulating valve and temperature sensor. Indicate how the controller set point and proportional band may be adjusted by the user.

(b) The steady-state heat-loss rate from a room is 240 W per degree temperature difference, inside to outside. The inside temperature is controlled by a proportional controller whose set point is 20°C and proportional band is 4°C. The heat output when the control valve is 100% open is 4800 W.

(i) Find the value of the steady-state inside temperature when the outside temperature is 10°C.

(ii) Find the minimum outside temperature at which the control valve will be just fully closed. (UMIST)

70. A frequency response test was carried out on an experimental room. The heater was of 1500 W maximum output and it had a negligible time constant. The amplitude and phase of the room-temperature response, when the heater was modulated sinusoidally at various frequencies are given in the table below.

(a) Find, graphically or otherwise, the transfer function relating heat supply to temperature response.

(b) The control loop is to be closed by a sensor of time constant 50 seconds and a proportional controller. Find a suitable value of proportional gain to give a *gain margin* of 6 dB.

Frequency (radian/s)	0.001	0.002	0.005	0.01	0.02	0.05	0.1	0.2
Amplitude (°C)	23.7	22.7	17.6	9.84	3.55	0.65	0.17	-
Phase (degrees)	-13	-27	-62	-100	-135	-161	-171	-175

(UMIST)

71. Define the *describing function* of a non-linear component and show how it may be used to predict the frequency and amplitude of oscillations in relay-controlled systems.

Given that the Fourier series for a square wave of amplitude M and frequency ω is

$$\frac{2M}{\pi} \sum_{n=1}^{\infty} \frac{1}{n} \sin n\omega t \ (n \text{ odd})$$

develop the describing function for a relay with total hysteresis h.

Use this function to estimate the frequency and amplitude of steady oscillations when a relay thermostat, with 3°C total hysteresis, is used to control the temperature of a room. The relay output can be taken as unity and the overall forward transfer function is

$$\frac{20\ \exp(-100s)}{1 + 1000s}$$

The feedback transfer function is unity. (UMIST)

72. Figure P.20 shows a shell and tube heat exchanger. Develop the transfer functions relating the secondary water outlet temperature θ_2 to the secondary water-inlet temperature θ_1, and to the temperature of the primary tubes θ_p. If a proportional controller is fitted to provide control of the secondary water-outlet temperature, obtain the equations for the time response of the controlled variable to unit step changes in (a) the reference variable, and (b) the disturbance variable.

By means of sketches illustrate the effect of controller gain on the form and magnitude of these time responses.

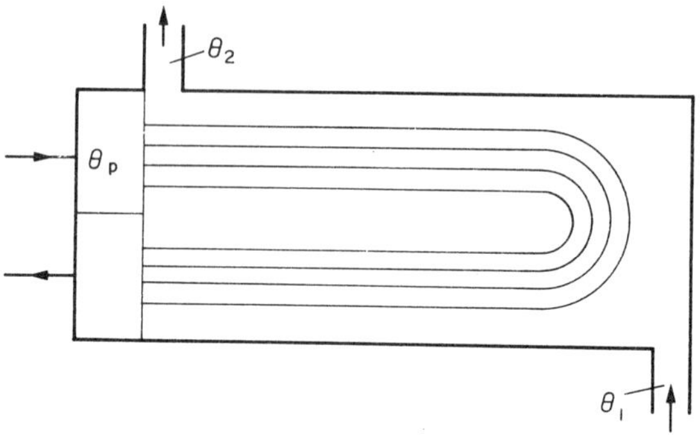

Fig. P.20 (UMIST)

73. A frequency response test was carried out on a model room. The heater was of 1500 W maximum output and it had a negligible time constant. The amplitude and phase of the room-temperature response, when the heater was modulated sinusoidally at various frequencies, are given below.

Deduce, graphically or otherwise, the transfer function connecting the heat output with the temperature response. The control loop is to be closed by a sensor of negligible time constant and a Proportional plus Integral controller. If the Integral Action Time is to be set equal to half the largest system time constant, find a suitable value of proportional gain.

Frequency radian/s	0.0005	0.001	0.002	0.005	0.01	0.02	0.05	0.1	0.5
Amplitude °C	74.2	72.1	64.6	37.1	14.4	3.80	0.37	0.037	-
Phase degrees	-6.3	-12	-24	-51	-75	-98	-129	-151	-174

(UMIST)

74. Develop Tsypkin's method for the analysis of thermostatically controlled systems.

The block diagram in Fig. P.21 represents an electric heater controlled by a room thermostat. Apply the method of Tsypkin to find the period of the controlled temperature cycle if the total hysteresis of the thermostat is 3°C. Also estimate the amplitude of the temperature cycle. (UMIST)

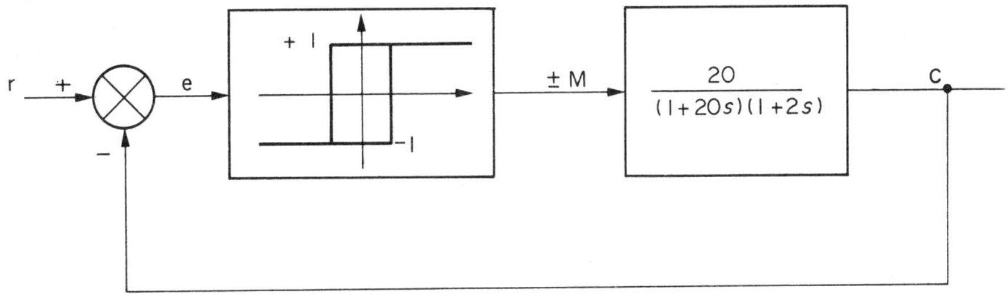

Fig. P.21.

APPENDIX

DERIVATION OF THE INVERSIONS OF SOME LAPLACE FUNCTIONS WHICH MAY BE USEFUL FOR STEP RESPONSE MODELLING

A1. Step Response of n First-order Non-interacting Elements in Series

$$F(s) = \frac{A}{s(1+a)^n} \quad \text{or} \quad \frac{A}{s(1+Ts)^n}$$

By Heaviside's expansion theorem the partial fraction expansion is as follows (see Section 5.3.1):

$$\frac{1}{s(1+a)^n} = \frac{Ao}{s} + \frac{Aan}{(s+a)^n} + \frac{Aan\text{-}1}{(s+a)^{n-1}} + \ldots$$

$$+ \frac{Aam}{(s+a)^m} + \ldots + \frac{Aa1}{(s+a)}$$

where $$Ao = \left[\frac{1}{(s+a)^n}\right]_{s=0} = 1/a^n$$

and $$Aam = \frac{1}{(n-m)!}\frac{d^{n-m}}{ds^{n-m}}\left[\frac{1}{s}\right]_{s\,=\,-a}$$

in general $$\frac{d^r}{ds^r}\left[\frac{1}{s}\right] = \frac{(-1)^r\ r!}{s^{r+1}}$$

Hence $$Aam = \left[\frac{(-1)^{n-m}}{s^{n-m+1}}\right]_{s\,=\,-a} = -a^{m-n-1}$$

From Table 5.1, $$L^{-1}\frac{Ao}{s} = Ao$$

and

$$L^{-1}\frac{Aam}{(s+a)^m} = \frac{t^{m-1}\,e^{-at}}{(m-1)!}\ ,$$

combining these results:

$$f(t) = \frac{1}{a^n} - e^{-at}\sum_{m=1}^{n}\frac{t^{m-1}\,a^{m-n-1}}{(m-1)!}\ .$$

Now letting $a = 1/T$ and dividing by T^n gives

$$F(s) = A/s(1+Ts)^n\ ,$$

and

$$f(t) = 1 - e^{-t/T}\sum_{m=1}^{n}\left(\frac{t}{T}\right)^{m-1}\frac{1}{(m-n)!}\ .$$

A2. Step Response of n First-order Elements (time constant T_2) with one First-order Element (time constant T_1), in Series

$$F(s) = \frac{1}{s(1+T_1 s)(1+T_2 s)^n}\quad\text{or}\quad\frac{1}{s(s+a)(s+b)^n}\ .$$

Expanding by Heaviside:

$$\frac{1}{s(s+a)(s+b)^n} = \frac{Ao}{s} + \frac{Aa}{s+a} + \frac{Abn}{(s+b)^n} + \frac{Abn-1}{(s+b)^{n-1}} + \ldots$$

$$+ \frac{Abm}{(s+b)^m} + \ldots + \frac{Ab1}{(s+b)}$$

$$= Ao/s + Aa/(s+a) + \sum_{m=1}^{n} Abm/(s+b)^m$$

where

$$Ao = 1/ab^n,$$

$$Aa = -1/a(b-a)^n,$$

$$Abm = \frac{1}{(n-m)!}\frac{d^{\,n-m}}{ds^{n-m}}\left[\frac{1}{s(s+a)}\right]_{s\,=\,-\,b}$$

$$= (-1)^{n-m}\sum_{i=1}^{n-m+1}\frac{1}{(-b)^i(a-b)^{n-m+2-i}}\ .$$

Inverting:

$$Ao/s \rightarrow 1/ab^n$$

$$Aa/(s+a) \rightarrow -e^{-at}/a(b-a)^n$$

and

$$\sum_{m=1}^{n}\frac{Abm}{(s+b)^m} \rightarrow e^{-bt}\sum_{m=1}^{n}\frac{Abm\ t^{m-1}}{(m-1)!}$$

collecting terms:

$$f(t) = L^{-1} \frac{1}{s(s+a)(s+b)^n}$$

$$= 1/ab^n - e^{-at}/a(b-a)^n + e^{-bt} \sum_{m=1}^{n} \frac{t^{m-1}(-1)^{n-m}}{(m-1)!} \sum_{i=1}^{n-m+1} \frac{1}{(-b)^i(a-b)^{n-m+2-i}} .$$

Now letting $a = 1/T_1$, $b = 1/T_2$ and dividing by T_1T_2 to give a unity numerator:

$$f(t) = L^{-1} \frac{1}{s(1+T_1s)(1+T_2s)^n}$$

$$= 1 - \left(\frac{T_1}{T_1-T_2}\right)^n \exp(-t/T_1)+\exp(-t/T_2) \sum_{m=1}^{n} \frac{(-1)^{n-m}t^{m-1}T_2{}^{2-m}}{(m-1)!} ,$$

$$\sum_{i=1}^{n-m+1} \frac{(-1)^i T_1{}^{n-m+1-i}}{(T_2-T_1)^{n-m+2-i}}$$

A3. Step Response of n First-order Elements (time constant T_2) with Two First-order Elements (time constant T_1) in series

$$F(s) = \frac{1}{s(s+a)^2(s+b)^n} .$$

Expanding by partial fractions as before:

$$F(s) = Ao/s + Aa2/(s+a)^2 + Aa1/(s+a) + \sum_{m=1}^{n} Abm/(s+b)^m$$

where $Ao = 1/a^2b^n$,

$Aa2 = -1/a(b-a)^n$,

$$Aa1 = \frac{n}{a(b-1)^{n+1}} - \frac{1}{a^2(b-a)^n} ,$$

$$Abm = \frac{(-1)^{n-m}}{(n-m)!} \sum_{i=1}^{n-m+1} \frac{(n-m+1)! - (i-1)(\overline{n-m})!}{(-b)^i(a-b)^{n-m+3-i}}$$

collecting time response terms as before:

$$f(t) = L^{-1} \frac{1}{s(s+a)^2(s+b)^n}$$

$$= 1/ab^n + e^{-at}\left(\frac{n}{a(b-a)^{n+1}} - \frac{1}{a^2(b-a)^n}\right) - \frac{t\,e^{-at}}{a(b-a)^n} +$$

$$+ e^{-bt} \sum_{m=1}^{n} \frac{(-1)^{n-m} t^{m-1}}{(m=1)!(n-m)!} \sum_{i=1}^{n-m+1} \frac{(n-m+1)! - (i-1)(n-m)!}{(-b)^{i} (a-b)^{n-m+3-1}}$$

and letting $a = 1/T_1$, $b = 1/T_2$ and dividing by $T_1^{\,2}\, T_2^{\,n}$;

$$f(t) = L^{-1} \frac{1}{s(1+T_1 s)^2 (1+T_2 s)^n}$$

$$= 1 + \left[\frac{T_1^{\,n}\, T_2}{(T_1-T_2)^{n+1}} - \frac{T_1^{\,n}}{(T_1-T_2)^n}\right] e^{-t/T_1} - \frac{T_1^{\,n-1}}{(T_1-T_2)^n}\, t\, e^{-t/T_1} +$$

$$+\exp(-t/T_m) \sum_{m=1}^{n} \frac{(-1)^{n-m} t^{n-m} T_2^{\,3-m}}{(m-1)!(n-m)!} \sum_{i=1}^{n-m+1} \frac{(n-m+1)! - (i-1)(n-m)!(-1)^{i} T_i^{\,n-m+1-i}}{(T_2-T_1)^{n-m+3-i}}$$

A4. Step Response of n First-order Elements in Series, all with Different Time Constants

$$F(s) = (s(1+T_1 s)(1+T_2 s) \ldots (1+Tms) \ldots (1+Tns))^{-1}$$

Expanding into partial fractions:

$$F(s) = Ao/s + A_1/(1+T_1 s) + A2/(1+T_2 s) + \ldots + Am/(1+Tms) + \ldots + An/(1+Tns)$$

$$= Ao/s + \sum_{m=1}^{n} \frac{Am}{(1+Tms)},$$

$Ao = 1$

and

$$Am = \frac{Tm^n}{\prod_{\substack{j=1\\ j\neq m}}^{j=n} (Tj-Tm)}$$

Hence $$f(t) = 1 - \sum_{m=1}^{n} \left[\exp(-t/Tm) \prod_{\substack{j=1\\ j\neq m}}^{j=n-1} \frac{Tm}{(Tj-Tm)}\right].$$

INDEX